The Terrestrial Macroinvertebrates of the Sub-Antarctic Îles Kerguelen and Île de la Possession

Series Editor
Marie-Christine Maurel

The Terrestrial Macroinvertebrates of the Sub-Antarctic Îles Kerguelen and Île de la Possession

Maurice Hullé
Philippe Vernon

with the collaboration of

Bernard Chaubet
Damien Fourcy
Romain Georges
Christelle Buchard

ISTE

WILEY

First published 2021 in Great Britain and the United States by ISTE Ltd and John Wiley & Sons, Inc.

ISTE Ltd
27-37 St George's Road
London SW19 4EU
UK

www.iste.co.uk

John Wiley & Sons, Inc.
111 River Street
Hoboken, NJ 07030
USA

www.wiley.com

Library of Congress Control Number: 2021937387

British Library Cataloguing-in-Publication Data
A CIP record for this book is available from the British Library
ISBN 978-1-78630-760-6

Photography: Bernard Chaubet (INRAE)
Cartography: Damien Fourcy (INRAE)
Species identification: Romain Georges (University of Rennes 1), Christelle Buchard (INRAE)

Contents

Foreword 1

The extraordinary American conservationist and thinker Aldo Leopold wrote: "To keep every cog and wheel is the first precaution of intelligent tinkering." To keep every cog and wheel presupposes that each one of them is not only known but can also be identified. For most places, knowing the full diversity of life is only imaginable. And for most people, identifying that diversity is unthinkable. Yet, a few specially endowed naturalists seem to be able to do so. And because they are often too little valued in our modern scientific world, their knowledge passes with them into history. Thus, places and the people who value them frequently fall into a cycle of knowing and forgetting.

Yet, every now and then, naturalists and writers come along who are determined to break that cycle. They recruit, through a combination of tenacity, charm and a fierce love for a place and its diversity, a host of helpers. Spurred on by the positive feedback that comes from enthusiasm, they recruit even more extensively. Eventually, they deliver a monograph of life. A record of a place's diversity and the means for others to know it in all of its exquisite detail. In doing so, they make forever known a part of our world that would otherwise have remained trapped in that cycle of light and dark.

Here, Maurice Hullé and Philippe Vernon have done just this for the larger invertebrates of the extraordinary sub-Antarctic islands of the Îles Kerguelen and the Île de la Possession.

Doing so could not have been easy. Scattered knowledge is one thing. But finding the animals to verify the knowledge and to captivate the attention of new recruits is quite another. The sub-Antarctic islands are simply gorgeous and apparently teeming with life. Yet, most are no longer what they once were. On islands free of introduced species – such as rodents, cats, carnivorous beetles, herbivorous slugs and system-altering weeds – the abundance of invertebrates is extraordinary. Beetles, moths and other insects are simply everywhere, even in the apparently hostile habitats such as fellfields. The coastal zones no longer stand out as special because of their abundance of insects (notably some very strange flies). Yet on islands such as the Îles Kerguelen and the Île de la Possession, where introduced species have had much time to do their work, abundances, except of the introduced species and some coastal inhabitants, are very low. Low to the point of thinking that some species have gone forever. Yet, often they have not. It simply takes a great deal of work to find them. And a very special talent of combining relentless enthusiasm for hard field work and exceptional charm to keep people coming back in search of knowledge, to look at more specimens and answer more questions. Questions that can be remarkably uncomfortable to ensure that the knowledge is secure.

Such tenacity, a characteristic of the authors, results in a handsome reward, and this monograph of the larger terrestrial invertebrate life of these islands is just such a reward. How extraordinary to have this work. The question "I wonder what this is?" is answered readily. And with that question addressed, immediately others can be posed. What does it do? What is its history of abundance? What is its future? How do we mitigate further change?

The sub-Antarctic islands are a global treasure. Most are recognized as such through their listing as World Heritage Sites. Here, we have a further record of the treasures of two of the most important island groups in the sub-Antarctic. As far as we can tell, and new genetic techniques are on the cusp of revealing, much of the terrestrial invertebrate diversity of the Indian Ocean sub-Antarctic islands originated here. In this respect, a history of accomplished French researchers, starting with René Jeannel, has been vital. These islands lie at the center of the evolutionary drama that is the sub-Antarctic. A drama that has captivated the minds of some of biology's greatest thinkers. Knowing the islands' diversity and being able to recognize it places that drama and its actors within everyone's grasp. Being able to tell the carabid beetles *Amblystogenium minimum* from *Amblystogenium pacificum* is now as easy as telling a Macaroni from a Rockhopper penguin.

The drama is not played out. The sub-Antarctic treasures face much difficulty because of introduced species, the local impacts of global climate change and what is now being increasingly documented as a positive effect of climate change on invasive species. If the play is not to end a tragedy, new actors are required; new actors with new parts. Parts that require thoughtful conservation action based on compelling evidence. In delivering this monograph, Maurice Hullé and Philippe Vernon have changed the nature of the auditions. They have broadened the talent pool to everyone who has a love of life as we know it. And they have extended the audience too. Even those who have never come close to the theatres that are these islands can take part by having the islands' diversity revealed.

This monograph is an exceptional tribute to the way in which good science, strong support for it and personal tenacity, enthusiasm and humor can be combined to deliver a lasting legacy. I recommend that as a reader you now plunge yourself into the world of sub-Antarctic terrestrial life, a world that, once seen, will never leave you. Cogs and wheels have never looked this good.

Steven L. CHOWN
Professor, Monash University
President, Scientific Committee on Antarctic Research
Melbourne, Australia
September 2020

Foreword 2

When confronted with a book about invertebrates of sub-Antarctic islands, we might expect a boring list written in a 19th-Century fashion by museum experts about their identification… But the multidisciplinary state of the art of two modern field naturalists makes this book fascinating. Maurice Hullé and Philippe Vernon have indeed been remarkably successful in providing a story on biodiversity, with nice iconography, which started in 1830 and is still evolving today.

It is also nice that this book provides a historical view, a well-deserved tribute to those pioneers who first described the remarkable endemic larger invertebrates of the sub-Antarctic Îles Kerguelen and the Île de la Possession (Eaton, Waterhouse, Jeannel, Dreux, Voisin, etc.) and on their expeditions (Challenger, Volage, Gauss, etc.). It was indeed the time of the first discovery and description of species. And also the time when similarities of these invertebrates were found with the remote wildlife of Africa and America. It raised puzzling questions on the way the invertebrates were once established on the islands.

But in the 1970s and 1980s, there was a huge academic pressure on biologists worldwide to abandon such projects, obviously involving the collection of long-time series, for a more mechanistic approach in cellular and molecular biology. As a result, systematicians and field biologists disappeared from universities and research organizations. Physiology, once predominant when molecular biology emerged, became a science of the past. However, as a pocket of resistance, long-term programs on biodiversity received logistics and funding by the successive French organizations in charge of sub-Antarctic research (*Mission de Recherche des Terres*

Australes et Antarctiques Françaises (TAAF)[1], *Institut Français de la Recherche et de la Technologie* (IFRTP)[2] and *Institut Polaire Paul-Émile Victor*[3] (IPEV)). Thanks to the principal researchers, Paul Tréhen and Yves Frenot, and thanks to the support of the *Centre National de la Recherche Scientifique* (CNRS), particularly in the recruitment of new scientists, long-term series were obtained.

Technicians and engineers also played a major role in the treatment of field data and samples. It is not surprising that some of them contributed largely to the preparation of this book. In particular, the engineer Romain Georges and the technicians Christelle Buchard and Bernard Chaubet should be acknowledged. Thanks to them and to all the volunteers who, over the years, have been overwintering in Kerguelen and the Île de la Possession. Our knowledge on the biology of the larger invertebrates of Kerguelen and the Île de la Possession has considerably increased over the last decades, as shown in this book.

The focus was particularly on their geographical distribution, their ecology and population dynamics. We note that it is also then that it was perceived that the endemic fauna of the islands was not in a preserved sanctuary but had to cope with an increasing number of introduced species. Meteorological recordings made it possible to characterize the first changes in ambient temperature and rainfall, indicating climate changes. They also showed that even a small increase in ambient temperature and drop in rainfall could facilitate the influx of invasive species.

With such knowledge being securely documented by field work and long-term series, this book illustrates how powerful it is to now enter a mechanistic and functional approach. Ecophysiology is the key to determining the environmental conditions which enable or limit the ability of invertebrate newcomers to survive and establish themselves on the islands and for the endemic species to deal with such invaders. Complementing the understanding of physiological mechanisms of adaptation, genomics open new perspectives to decipher and understand the origin and biogeography of sub-Antarctic invertebrate fauna.

Thus, altogether, this book is not only about the listing of the larger invertebrates of Kerguelen and the Île de la Possession. It is a truly remarkable story on a field and laboratory research which is still ongoing, providing detailed information on

1 Research mission of the The French Southern and Antarctic Lands.
2 French Institute of Science and Technology.
3 The French Polar Institute Paul-Émile Victor.

the origin, evolution and ecosystem functioning of the invertebrate fauna of sub-Antarctic islands.

Yvon LE MAHO
Emeretus Director of Research, CNRS, Institut Pluridisciplinaire Hubert Curien (University of Strasbourg/CNRS) and Centre Scientifique de Monaco
Member, French Academy of Sciences and Chair of the board, French Polar Institute
Strasbourg, France
September 2020

Acknowledgements

This book is based on extensive fieldwork that was only possible thanks to the logistics and funding provided by the administration of the French Southern and Antarctic Territories (TAAF) and the French Polar Institute Paul-Émile Victor (IPEV).

We are grateful to the leaders of the successive research programs, dedicated to the study of biodiversity and the functioning of terrestrial ecosystems, where this work took place: Paul Tréhen, Yves Frenot, Marc Lebouvier and David Renault.

From the very first missions that followed the establishment of the bases to the present day, many explorations have been carried out in almost all sectors of the Îles Kerguelen and Île de la Possession. Many people have made it possible to reach an almost exhaustive knowledge of the macroinvertebrate fauna. This involves more than 150 people: all the young wintering field assistants and the scientists in summer campaigns, as well as the civilian and military staff dedicated to logistics and who often accompanied the scientists into the field. May we thank them all here collectively!

We would like to thank Marc Lebouvier who made a major contribution to data acquisition and management, and Christelle Buchard and Romain Georges, who have done a lot of work on species identification, collection management and the identification training we put in place each year for young wintering scientists. We are also very pleased to have been able to illustrate most of the invertebrate species present on the Îles Kerguelen and the Île de la Possession thanks to our fellow photographer, Bernard Chaubet. Damien Fourcy drew all the maps of the Kerguelen and Crozet islands specifically for this book.

We would also like to thank the specialists who brought their expertise on the chapters devoted to the different taxonomic groups: Yves Frenot (earthworms), Maryvonne Charrier (mollusks), Julien Pétillon and Cyril Courtial (spiders), Steven Chown (weevils), Vladimir Gusarov (rove beetles), Adam Broadley (sciarids), Torbjorn Ekrem and Aina Maerk Aspaas (chironomids), Evelyne Turpeau and Christelle Buchard (aphids) and Richard Harrington (lepidopterans).

We are grateful to the "Zone Atelier Antarctique et Terres Australes" (https://zaantarctique.org/) and to Françoise le Moal for editing the meteorological data from the Kerguelen and Crozet weather stations provided by Météo France. We would also like to thank Clément Quetel and Floran Hoarau of the French Southern Lands National Nature Reserve for the information on the reserve and the maps of the access areas, as well as Yann Laurent and Franck Duval for their drawings of insect morphology.

David Hatcher and Richard Harrington provided invaluable help in proofreading and correcting our French-English.

The entire ISTE team were also a great help in the conception of the book.

Thanks to our supervisory institutes, the University of Rennes 1, the National Centre for Scientific Research (CNRS) and the National Research Institute for Agriculture, Food and Environment (INRAE) for hosting our research activities and signing our mission orders!

Finally, we express our warmest gratitude to Steven Chown and Yvon Le Maho, who agreed to write the forewords of this book and with whom we are happy to share our passion for the sub-Antarctic islands.

Introduction

The Îles Kerguelen and the Îles Crozet belong to the "Terres Australes et Antarctiques Françaises" (TAAF). They are located in the South Indian Ocean Province. They were discovered in 1772 respectively by Yves Joseph de Kerguelen de Trémarec and Marc-Joseph Marion Dufresne whose first mate, Julien Crozet, had landed on Île de la Possession (Delépine 1995).

After their discovery, these islands remained untouched by any human presence until the whalers and the sealers visited them during the 19th Century and the beginning of the 20th Century. In 1924, they were officially attached to France, who only installed permanent bases there in 1950 on the Îles Kerguelen (Port-aux-Français) and in 1963 on Île de la Possession (Base Alfred Faure).

The history of fauna and flora is affected by the history of human visits. Fauna and flora were originally composed of a small number of native sub-Antarctic species. The first introductions of alien species occurred as early as the 19th Century (rabbits and dandelion in 1874). The number of introduced species increased significantly from the mid-20th Century, and now makes up approximately half of the species present. Some of the introduced species have become invasive. Given the small number of native species and the quantitative importance of introductions, terrestrial ecosystems have been severely disrupted by these invasions, especially as global warming has also caused significant changes in vegetation cover.

Assessing the impact of introduced species on native ecosystems and deciphering the interactions between native and alien species are some of the issues currently being investigated by scientific teams.

Until now, there has been no work presenting all the invertebrate fauna of these islands. Here, we are only considering macroinvertebrates (earthworms, mollusks, spiders and insects) from the Îles Kerguelen and Île de la Possession for which we have an almost complete knowledge. Other groups such as springtails, mites and tardigrades are still undergoing a taxonomic revision.

In the first part, we present the general characteristics of the Îles Kerguelen and Île de la Possession: their geography, their climate and their main landscapes, which constitute the environment and the habitats of terrestrial invertebrates. We also dedicate a paragraph to the first scientific discoveries that marked the knowledge of the natural history of these islands, then briefly mention the current scientific themes concerning invertebrates, as well as the measures of protection that have been put in place since the creation of a natural reserve in 2006.

In the second part, we present all species of macroinvertebrates. For each one, we evoke its status: *endemic* for the species found only on one of the two archipelagos, *native* for those present on other islands of the South Indian Ocean Province or other sub-Antarctic provinces, *introduced* for naturalized alien species and finally *transient* for species observed regularly but who are not reproducing on the islands. We also give information on known distribution, identification and elements of ecology. For the Archipel Crozet, we only present the fauna of Île de la Possession because the other islands have not been sufficiently explored to have such a thorough knowledge of the species.

1

General Presentation

1.1. Physical environment

1.1.1. *Location and size*

The sub-Antarctic Îles Kerguelen and Îles Crozet belong to the South Indian Ocean Province, as do the Heard and McDonald Islands, and the Prince Edward Islands (Marion Island and Prince Edward Island) (Figure 1.1). Only Heard Island and McDonald Island are located in the very cold Antarctic waters. The others are located above the Polar Front, separating the very cold Antarctic waters from the sub-Antarctic cold waters.

The Îles Kerguelen are located at 49°21'S, 70°13'E (Port-aux-Français) and Île de la Possession at 46°24'S, 51°46'E (Base Alfred-Faure). They are respectively 3,250 km and 2,800 km from Île de la Réunion, 3,800 km and 2,400 km from South Africa and 1,900 km and 2,100 km from the Antarctic Continent. The Îles Kerguelen and Heard Island are located on the same continental shelf and are only 423 km apart. The Îles Crozet and the Prince Edward Islands are a second, more western subgroup.

The Îles Kerguelen are 130 km long and 120 km wide. Their total area is 7,215 km^2 including 6,675 km^2 for the main island which is called Grande Terre (Figure 1.2). The Îles Crozet are composed of an eastern group of islands (Île de l'Est and Île de la Possession) and a western group about 110 km away (Îlots des Apôtres, Île aux Cochons and Île des Pingouins) (Figure 1.3). Île de la Possession, the only inhabited island, is much smaller than Grande Terre, at only 150 km^2.

1.1.2. *Topography*

The morphology of the Îles Kerguelen is mainly related to their magmatic origin and glacial erosion. Their geological history is long. The most ancient parts emerged about 35–40 million years ago and the last magmatic events date back to about 30,000 years ago. Mount Ross rises to the South at 1,830 m a.s.l. Part of the western and central regions is covered by an ice cap, the Cook cap, which rises to 1,000 m a.s.l. and covers about 550 km². The North (Péninsule Loranchet) and the South-West (Péninsule Rallier du Baty) are mountainous and rise to 1,260 m a.s.l. To the east, Péninsule Courbet constitutes a vast area of glacial erosion which is at a low altitude and is dotted with numerous lakes (Giret *et al.* 2003a) (Figure 1.4).

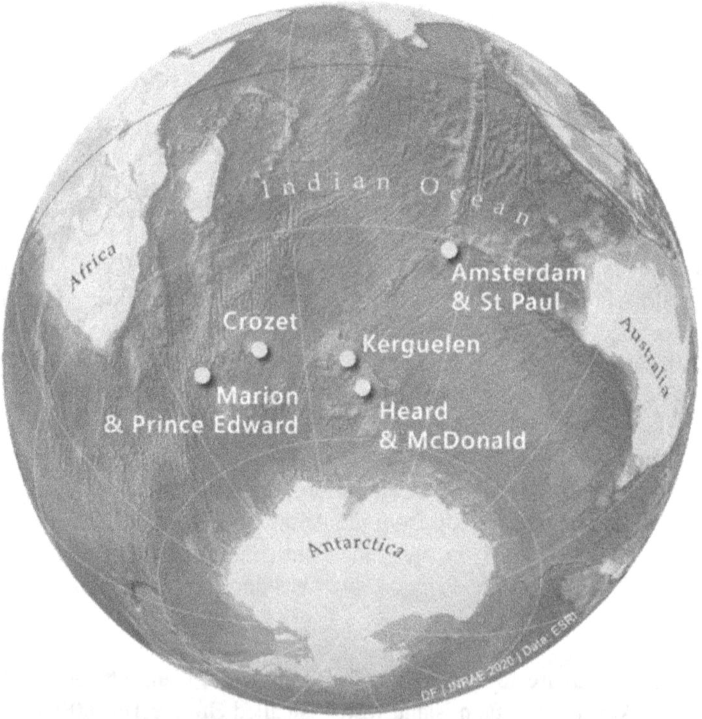

Figure 1.1. *South Indian Ocean sub-Antarctic islands*

Île de la Possession was formed between 8 million years ago and about 10,000 years ago for the most recent episodes (Delépine 1973; Giret *et al.* 2003b). The island culminates at 934 m a.s.l. at Pic du Mascarin. Large alluvial valleys (Vallée de la Hébé and Vallée des Branloires) open onto the north-east coast (Figure 1.5).

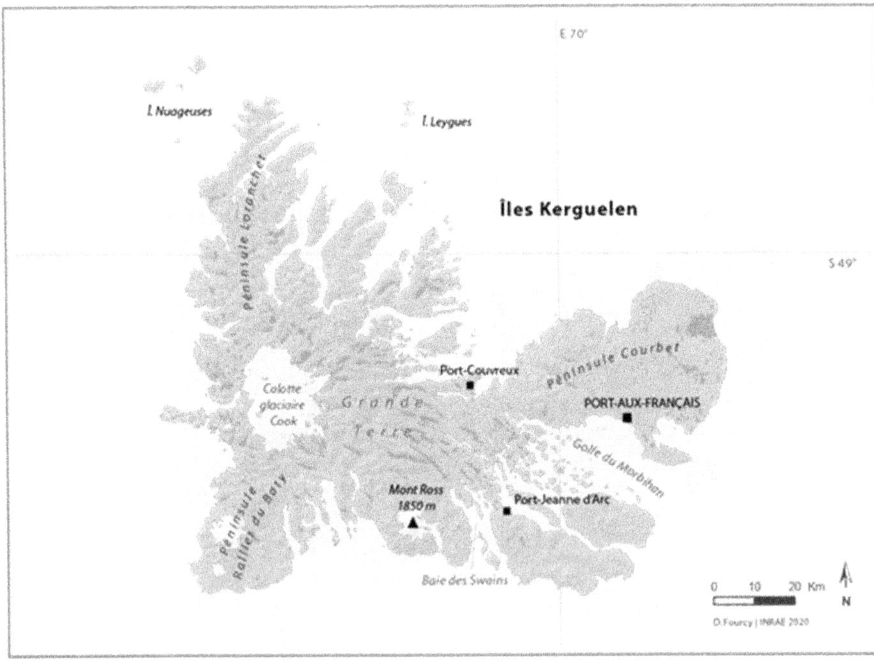

Figure 1.2. *The Îles Kerguelen*

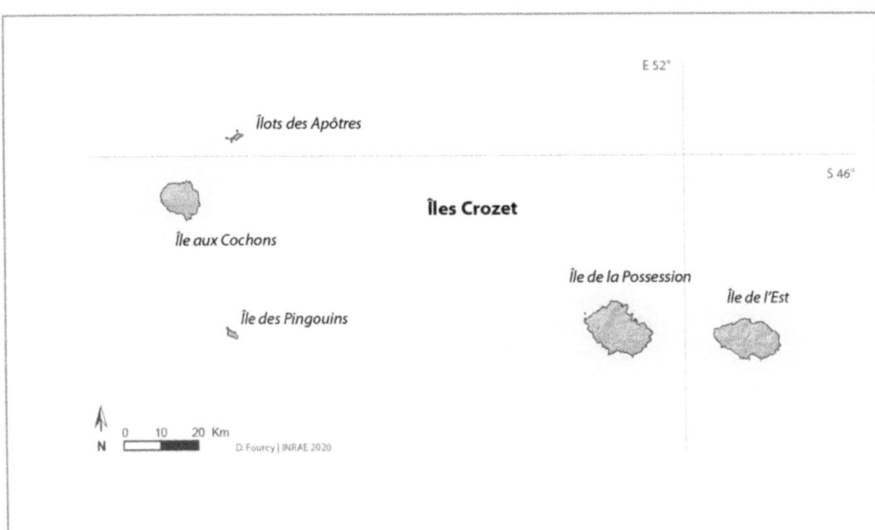

Figure 1.3. *The Îles Crozet*

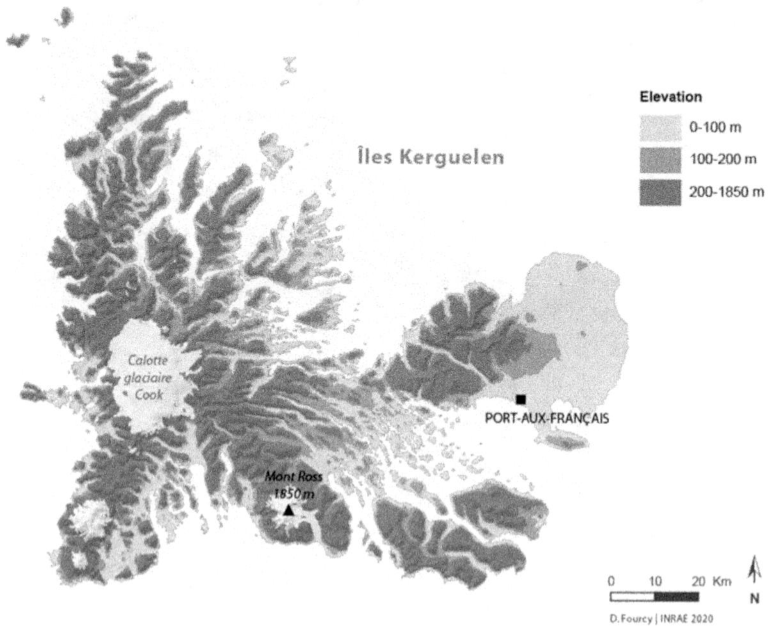

Figure 1.4. *Topography of the Îles Kerguelen*

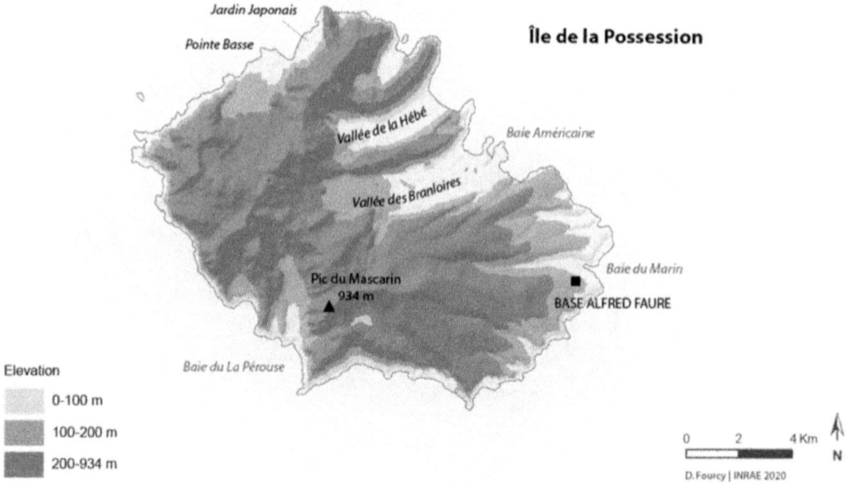

Figure 1.5. *Topography of Île de la Possession (Crozet)*

A large part of the two islands, Kerguelen and Possession, is mountainous with an altitude of over 200 m a.s.l. Areas below 100 m in altitude represent 32% and 22% respectively on the Îles Kerguelen and Île de la Possession, and areas below 200 m represent 50% and 40%, respectively (Figures 1.4 and 1.5).

1.1.3. *Climate*

As with all remote sub-Antarctic islands, both the Îles Kerguelen and Île de la Possession have an oceanic cold climate, strongly influenced by the South Indian Ocean. They are located at the same latitude as France and therefore have an equivalent photoperiodic regime, with long days in summer and short days in winter. The mean air temperature for 1951–2013 at sea level was 4.8°C at Îles Kerguelen and 5.5°C at Île de la Possession (Figure 1.6). Seasonal variations are low, between 2°C and 8°C at Îles Kerguelen and 3°C and 8°C at Île de la Possession. Summer temperatures may temporarily reach or exceed 15°C. At the Îles Kerguelen, average temperatures increased by 1.9°C between 1964 and 1981 and then stabilized to around 5.2°C. Concomitantly, the number of freezing days has decreased from 130 days per year in the 1960s to 100 days per year now. At Île de la Possession, the number of freezing days is less than 60 per year.

Air temperature decreases by approximately 1°C per 100 m increase in altitude (Pointe Suzanne and Port-Jeanne d'Arc, Kerguelen; M. Hullé unpublished data). This steep decay means that the thermal thresholds of development of many insects may be reached rapidly with altitude. The mean temperature is below 3°C from 200 m a.s.l., and insect life is consequently mainly concentrated on low-lying coastal areas (Figures 1.4 and 1.5).

The average rainfall of Île de la Possession is high with 2,390 mm per year (Figure 1.6). On the Îles Kerguelen, there is a rainfall gradient from west to east. The west coast receives about 2,000 mm per year, and Port-aux-Français, located to the east, receives about 700 mm per year. At Port-aux-Français, rainfall oscillates between years of drought, for example, 1965 with 350 mm, and wet years, for example, 1987 with 1,150 mm.

Winds are predominantly from the northwest, west and southwest. The average wind speed is 35 km/h on the Îles Kerguelen and 38 km/h on Île de la Possession. Strong winds are frequent. They are mostly from the northwest (Figure 1.7). The strongest gusts can exceed 200 km/h. They reach an average speed of 131 km/h on the Îles Kerguelen and 141 km/h on Île de la Possession. The number of days with winds above 100 km/h is 68 per year on the Îles Kerguelen.

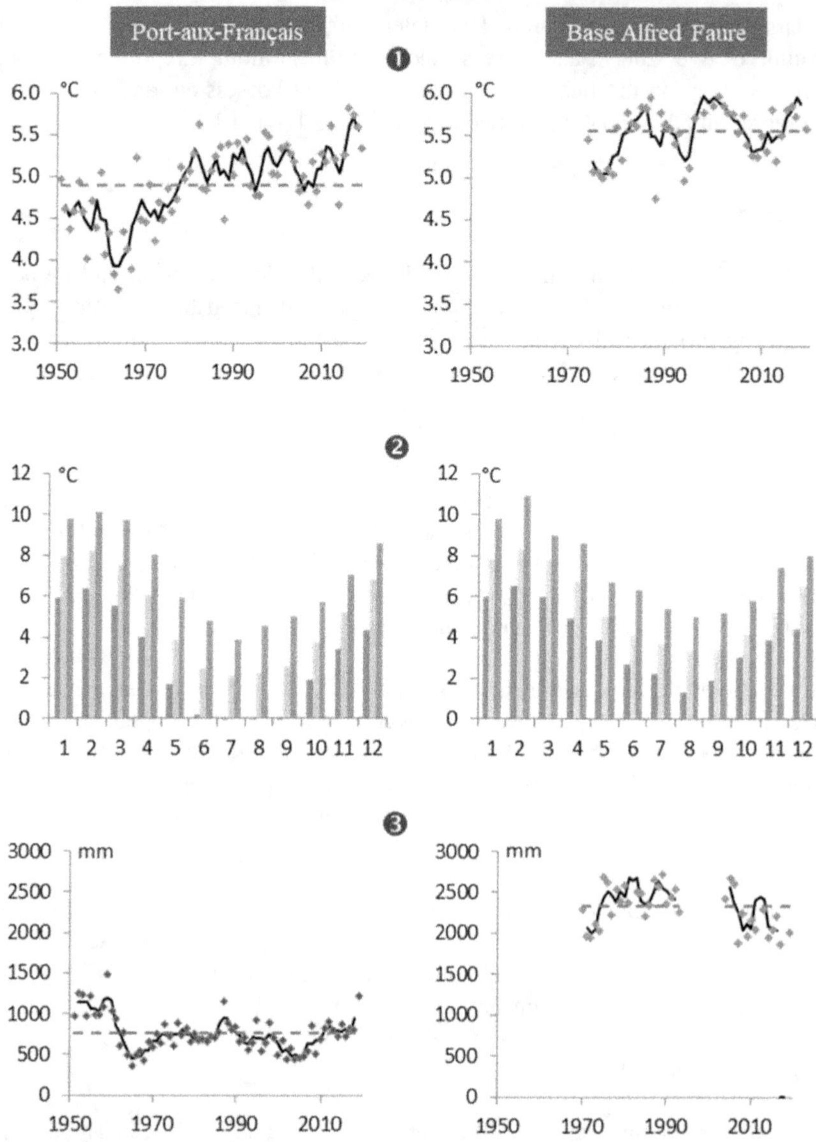

Figure 1.6. *Climate at Port-aux-Français (Îles Kerguelen) (left) and Base Alfred-Faure (Île de la Possession) (right): ❶ Annual mean temperature, ❷ Monthly temperature (mean in gray, maximum in red and minimum in blue), ❸ Annual precipitation (data from Météo France)*

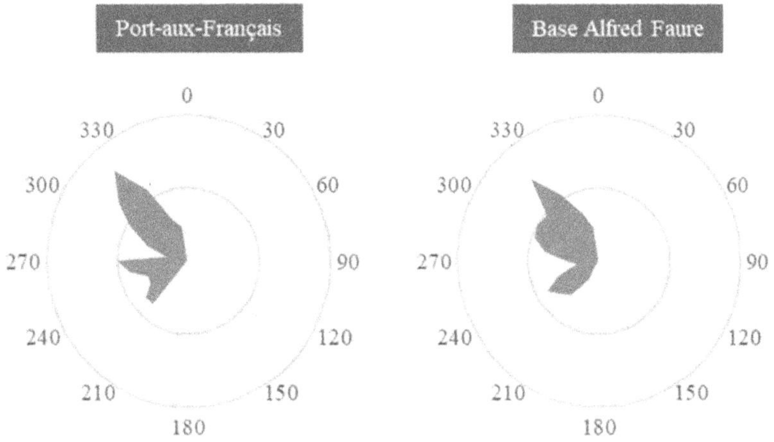

Figure 1.7. *Direction of the strongest winds (data from Météo-France)*

1.2. Landscapes

Macroinvertebrates live mainly in low-elevation areas, which present four major types of landscape: sea-edge, grassland, wetland/mire and fellfield/rock (Davies and Greene 1976). Buildings and huts are also a habitat for some invertebrates.

1.2.1. *Sea-edge*

The seaside habitats (Figure 1.8) may consist of bare rocky areas such as pebble beach, blocks or shore platforms. They can be also covered with *Crassula moschata* or with coastal meadows, where native plants such as *Leptinella plumosa*, *Pringlea antiscorbutica* or *Poa cookii* dominate. Introduced plants such as *Taraxacum officinale* and *Poa annua* can also be locally abundant. These habitats are under the direct influence of marine inputs that are stranded algae. This oceanic input constitutes a very important food resource for terrestrial fauna. It is also the area occupied by colonies of seabirds and marine mammals whose feces, carrions and broken eggs are a major source of food for invertebrates.

The small coastal cliffs that are covered with vegetation harbor a significant diversity of invertebrate species. These are probably the richest environment for macroinvertebrates (Figure 1.9).

Rocky coast

Degraded soil with *Crassula moschata*

Shore platform

Pebble beach with stranded algae

Pringlea antiscorbutica and *Leptinella plumosa*

Littoral grassland with *Aceana magellanica* and *Poa annua*

Figure 1.8. *Sea-edge habitats (photos: M. Hullé)*

Figure 1.9. *This littoral cliff at Pointe Suzanne (Îles Kerguelen), covered with* Pringlea antiscorbutica, Leptinella plumosa, Poa cookii *and* Poa annua, *was home to 18 different invertebrate species such as the spiders* Myro kerguelenensis, Neomaso antarcticus, *the moth* Embryonopsis halticella, *the beetles* Meropathus chuni, Bothrometopus brevis, Ectemnorhinus viridis, Merizodus soledadinus, Cartodere nodifer, *the flies* Amalopteryx maritima, Apetaenus litoralis, Anatalanta aptera, Calycopteryx moseleyi, Fucellia tergina, Psychoda parthenogenetica, *and the aphids* Rhopalosiphum padi *and* Myzus ascalonicus *(photo: M. Hullé)*

The seaside is the exclusive habitat of some species such as the weevil *Palirhoeus eatoni* and the flightless flies *Apetaneus enderleini* and *Paractora dreuxi*.

1.2.2. *Grasslands*

Grasslands (Figure 1.10) are present everywhere from the coast and up to 200 m a.s.l. These may be either almost monospecific grasslands with native plants such as *Acaena magellanica*, *Leptinella plumosa* or *Azorella selago* or introduced plants like grasses or dandelions. There are also mixed grasslands where native plants are in competition with one another or with introduced species. The monospecific meadows of *Leptinella plumosa* are coastal or supra-littoral, those of *Acaena magellanica* occupy the whole height and cover a large part of the slopes and those of *Azorella selago* are situated at higher altitudes and can develop up to 700 m a.s.l.

The meadows can be mixed with the rocky habitats. All the intermediaries between soil completely covered with vegetation and bare soil are possible. In the same way, grasslands may be very wet or very dry with plants in water deficit. Finally, coastal meadows can be heavily trampled by marine animals.

The invertebrate fauna of grasslands depend on plant composition, the proportion of area covered by vegetation or rocks and the water status of the soil. Introduced species such as the spider *Tenuiphantes tenuis* and the aphid *Myzus ascalonicus* may be plentiful. There is also the small native spider, *Crozetulus minutus*.

1.2.3. *Wetlands and mires*

The high humidity and rainfall, especially on Île de la Possession and the western part of the Îles Kerguelen, lead to the presence of numerous wetlands, rivers and lakes (Figure 1.11). Valley bottoms and slopes can be particularly soggy and oozing, making walking difficult. Native plants like *Acaena magellanica*, *Agrostis magellanica*, *Callitriche antarctica*, *Uncinia compacta*, *Juncus scheuchzerioides* and *Ranunculus biternatus* easily cope with this soil moisture. There are also bryophytes and liverworts.

These areas are the domain of some earthworms, spiders, Diptera like chironomids and black flies, and some aphids.

1.2.4. *Fellfield and rocks*

Fellfield is a widespread habitat from 100 m a.s.l. and occasionally below (Figure 1.12). It can be more or less drained and covered with vegetation such as the cushion plant *Azorella selago* and grasses like *Poa kerguelensis*. There are some weevils of the genus *Bothrometopus* and *Ectemnorhinus*, moths of the genus *Pringlea* and spiders of the genus *Myro*. It is the exclusive habitat of certain species such as the endemic sting bug *Phthirocoris antarcticus*. Rock walls, scree and rocky blocks are also habitats for some invertebrates including weevils and spiders.

Acaena magellanica

Mixed grassland with native and introduced plants

Slopes covered with *Acaena magellanica*

Slopes covered with *Poa pratensis*

Grassland overground with dandelion

Mixed grassland

Figure 1.10. *Grasslands (photos: M. Hullé)*

Wet valley background

River banks with *Acaena magellanica*

Mire covered with *A. magellanica*

Slopes covered with oozing *A. magellanica*

Slope covered with bryophyte

Elephant seal resting places

Figure 1.11. *Wetlands and mires (photos: M. Hullé)*

Xeric fellfield with cushion of *Azorella selago* | Xeric fellfield

Mesic fellfield | Mineral soil

Rocky walls | Mineral soil subjected to frost

Figure 1.12. *Fellfield and rocky habitats (photos: M. Hullé)*

Port-aux-Français (Kerguelen)

Base Alfred Faure (Crozet)

Port-Couvreux (Kerguelen)

Port Jeanne d'Arc (Kerguelen)

Île Longue (Kerguelen)

Pointe Basse (Crozet) (photo R. Joly)

Figure 1.13. *Buildings and huts (photos: M. Hullé, unless otherwise stated)*

1.2.5. *Buildings and huts*

Base buildings and field huts are important habitats, especially for introduced species (Figure 1.13). These are, for example, flies such as *Fucellia tergina* and those belonging to the family *Carnidae*, and spiders such as *Tegenaria domestica, Steatoda grossa* and *S. triangulosa*.

The lands in the immediate vicinity of the bases host more introduced species than the more distant lands. The bases, which concentrate human activity and receive most newcomers, are the main introduction points of non-native species that need to acclimatize there before spreading.

All greenhouses were decommissioned in 2009 following the establishment of a national nature reserve but, in the past, these have probably played an important role in the acclimation phase of some introduced species. This was without doubt the case for aphids which, today, are among the most invasive species.

1.3. Human occupancy

The belief in a southern land as a counterbalance and necessary symmetry of the northern hemisphere is ancient. Eighteenth-Century scholars pushed to discover this southern continent and provoked some great exploratory expeditions. Commercial companies such as the "Compagnie des Indes" also encouraged these expeditions to search for havens other than Île de France (Mauritius) on the way to India, as well as to the Atlantic. Jean-Baptiste Bouvet-Lozier, Marc-Joseph Marion-Dufresne and Julien Crozet were captains of the "Compagnie des Indes", while Louis-Antoine de Bougainville and Yves-Joseph de Kerguelen were officers in the service of the King of France.

The French sub-Antarctic islands were discovered in 1772 during these expeditions: the Îles Crozet by Marc-Joseph Marion-Dufresne and Julien Crozet, who landed on Île de la Possession at Crique du Navire and the Îles Kerguelen by Yves-Joseph de Kerguelen (this is Charles de Boisguehenneuc, the second captain, who took possession of the island at Anse du Gros-Ventre).

All these islands were revisited four years later by James Cook who spent Christmas 1776 in Port-Christmas in the North of Kerguelen. On this occasion, they described Kerguelen's cabbage, *Pringlea antiscorbutica*, thus producing what was probably the first scientific publication concerning these islands. Cook published his voyage in 1784, which made the positions and descriptions of the islands known and led very quickly, as early as 1792, to the first American whaling and sealing expeditions (Delépine 1995).

The 19th Century was one of intensive exploitation of marine mammals and seabirds, with more than 1,200,000 sea lions slaughtered for their oil, as well as whales, sea furs and penguins (Basberg and Headland 2008). The number of ships was extremely important: up to 700 per year according to James Clark Ross, who visited the islands in 1840 (Ross 1847).

France again took possession of the Îles Kerguelen in 1893 and therefore had to ensure a permanent presence on the islands. The country gave a concession to René Emile Bossière and his brother Henry Bossière for 50 years. The two brothers developed several projects such as a whaling factory in Port-Jeanne-d'Arc (1908–1926) and a sheep farm in Port-Couvreux (a first period in 1913 but stopped by World War I and a second period between 1922 and 1931). On this occasion, they introduced forage plants along with a predatory insect, *Merizodus soledadinus*. The French state also took possession of the Îles Crozet in 1931.

Several military ships frequented the southern islands during World War II. Some wrecks and cemeteries still bear witness to this.

The first permanent bases were installed in 1950 on the Îles Kerguelen (Port-aux-Français, Grande Terre) and 1963 on Île de la Possession (Base Alfred-Faure). These bases regularly host between 25 and 50 people in base Alfred-Faure and from 50 to 100 in Port-aux-Français. Bases are refueled by boat four times a year and are also visited by fishing boats several times a year. Boat is the only means of access. In addition to the technical, military and scientific staff, it brings 60 tourists every year who enjoy the trip but who do not stay on the islands except during the stopovers.

1.4. Natural history and history of species discovery

Our current knowledge of native and introduced invertebrate species is based on the first scientific expeditions that took place in the second half of the 19th Century and the first half of the 20th Century (Box 1.1) and on the scientific programs that were initiated after the bases were settled.

The first species were brought back by the expeditions of the *Erebus* and the *Terror* in 1843, and the first descriptions were the weevil *Ectemnorhinus viridis* and the snail *Notodiskus hookeri* in 1853 and 1854, respectively.

Thirty years later, the very successful expedition of the *Challenger* allowed the description of 14 new species, mainly Diptera by Alfred Edwin Eaton and weevils by Charles Owen Waterhouse. At the end of the 19th Century and the beginning of the 20th Century, the expeditions of *Valdivia* (1898) and *Gauss* (1902) allowed the description of 18 new species, particularly by Günther Enderlein who described 15 native species including two moths of the genus *Pringleophaga.* The pre-war expeditions of Edgar and Andrée Aubert de la Rüe and René Jeannel allowed the description of spiders by Vernon Hickman and weevils by René Jeannel.

Thus, the discovery of 66 native species occurred over 140 years, from 1854 to 1994 (Figures 1.14 and 1.15).

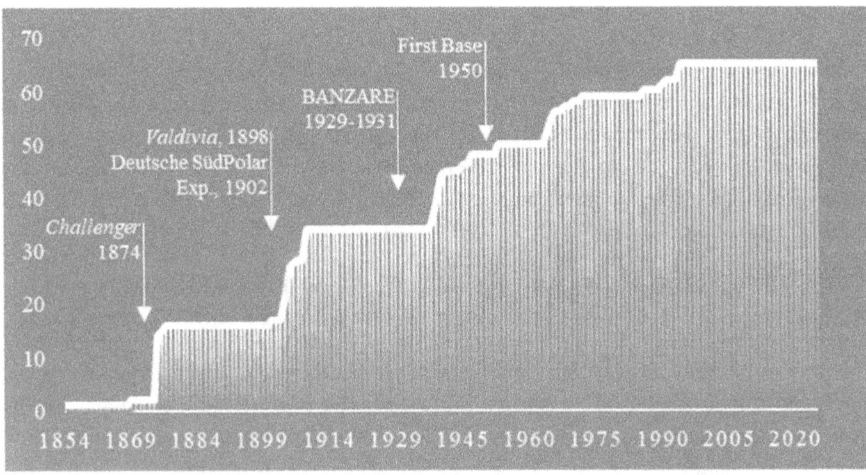

Figure 1.14. *Date and cumulated number of native species' first description (cumulative data for the Îles Kerguelen and Île de la Possession)*

More curiously, a few alien species have also been described from specimens found on the islands, even though these species may be common in their areas of origin. This was the case, for example, of the ant *Camponotus werthi* discovered by Emil Werth in 1902 (Deutsche Südpolar Expedition) and described by Auguste Forel (1908). This ant came from Cape Town in South Africa. Similarly, the book louse, *Rhyopsocus eclipticus*, described by Hermann August Hagen (1876), was discovered in Baie de l'Observatoire (Kerguelen) by the American expedition coming from Washington (USA), to observe the transit of Venus in 1874. It was probably introduced with the packing straws used on board the *Swatara* to protect the optical material. Similarly, the genus *Limnophyes* was created by Alfred Edwin Eaton in 1875 to describe the species *L. pusillus*, which was later renamed *L. minimus*, previously described by Johann Wilheim Meigen in 1818.

The 19th Century was the century of several important scientific expeditions. Scientists and hunters have come together for more than 150 years. The main scientific expeditions were as follows (Headland 1989; Delépine 1995):

1839–1843: the *Erebus* and the *Terror* commanded by James Clark Ross and Francis Crozier. Their main mission was the search for the magnetic pole, but naturalists like Robert McCormick, Joseph Dalton Hooker and David Lyall were also on board.

1874: the *Challenger* commanded by Georges Nares and whose chief scientist was Charles Wyville Thomson, a marine biologist, accompanied by the naturalists and zoologists John Murray, Rudolf von Willemoes-Suhm and Henry Nottidge Moseley.

1874: the *Volage*, the *Supply*, the *Swatara*, the *Gazelle* and the *Monongahela* to observe the transit of Venus. The astronomer Stephen Joseph Perry and the entomologist Alfred Edwin Eaton were present. These expeditions did not have only positive consequences since the first rabbits were released on Grande Terre (Îles Kerguelen) by the people aboard the *Volage* and the first dandelion seeds sown by those aboard the *Swatara* to provide food. This was the source of the first ecological disasters.

1898–1899: the *Valdivia* with the oceanographer Carl Chun as expedition leader, the botanist Wilhelm Schimper and the zoologist Ernst Vanhöffen on board.

1901–1903: the *Gauss* and the *Tanglin* with the geologist Erich von Drygalski as the outstanding "Deutsche Südpolar Expedition" leader. His goal was the Antarctic, but he organized a one-year stay on Kerguelen for five people including the scientists Emil Werth, Josef Ezensperger and Karl Luyken, and Ernst Vanhöffen was also on board.

1908–1909 and 1913–1914: the *J.B Charcot* and then the *Curieuse* led by brothers Raymond and Henri Rallier du Baty who drew the first good map of the interior of Kerguelen and made many observations of the flora and the fauna.

1924: the *Oural* with Etienne Peau, curator of Le Havre National History Museum, at Îles Kerguelen.

1928, 1931, 1949 and 1952: with Edgar and Andrée Aubert de la Rüe who came several times, essentially for geological studies.

1929–1931: the *Discovery* stopped at the Îles Crozet, the Îles Kerguelen and Heard Island before continuing south to Antarctica. This British, Australian and New Zealand Antarctic Research Expedition (BANZARE) was under the command of Douglas Mawson.

1938–1939: the *Bougainville* commanded by Henri Fabre de la Ripelle and with the entomologist René Jeannel on board. This expedition passed by Marion Island, the Îles Crozet and the Îles Kerguelen.

Box 1.1. *Main scientific expeditions of the 19th and early 20th Centuries*

Since the installation of the bases, scientific missions have been able to take place each year with stays of several months, first thanks to the "Mission de Recherche des Terres Australes et Antarctiques Françaises", which became the "Institut Français pour la Recherche et la Technologie Polaires" (IFRTP) in 1992, and finally, the "Institut Polaire Français Paul-Émile Victor" (IPEV) since 2002. Fifteen new native species have been described since then. Many scientists have succeeded each other in the field. They collected many samples, which they identified themselves or entrusted the identification to specialized colleagues.

These expeditions also made it possible to draw up a complete list of the introduced species and to follow the history of their acclimation and expansion. This was the case, for example, for the ground beetle *Merizodus soledadinus* introduced at Port-Couvreux, Îles Kerguelen, in 1913, systematically researched by René Jeannel in 1939 and then by the different entomologists that have followed (see Figure 5.38).

The toponymy of the sub-Antarctic Îles Kerguelen and Îles Crozet (Delépine 1973) pays tribute to some of these pioneering scientists for invertebrate research: Lac Eaton, Péninsule Gauss, Cap Jeannel, Mont Moseley, Mont Etienne Peau, Port Perrier, Rivière Studer, Île Suhm, Bras Vanhöffen, Mont Werth and Mont Wyville Thomson are examples.

1.5. Current research on invertebrates

The study of macroinvertebrates is the subject of the research program "Subanteco" supported by the "Institut Polaire Français Paul-Émile Victor" (IPEV). This program seeks to evaluate the impact of environmental changes and biological invasions on native communities and more generally, on the functioning of terrestrial ecosystems. This program is based on the long-term monitoring of changes in the distribution of native and introduced species and on experimental designs, in order to respond to three main questions: 1) how much are distribution patterns and community structure of terrestrial fauna and flora changing over time? 2) What are the effects of changing environments and multiple stresses on species physiological ecology? 3) What are the dynamics of invasive species, and how much are they affecting sub-Antarctic biodiversity? This makes it possible to better understand the adaptive responses of species to changes and to better understand the processes underlying biological invasions. It also helps decision-making for those responsible for the management and development of these areas.

Joseph Dalton Hooker
(1817-1911)

Charles Wywile Thomson
(1830-1882)

Günther Enderlein
(1875-1968)

Erich Dagobert von Drygalski
(1865-1949)

René Jeannel
(1879-1965)

Andrée and Edgar Aubert de la Rüe
(1903-1990 / 1901-1991)

Charles Owen Waterhouse
(1843-1917)

Eugène Séguy
(1890-1985)

Vernon Victor Hickman
(1894-1984)

Figure 1.15. *Portraits of some naturalists who worked on the invertebrate fauna of the Îles Kerguelen and the Îles Crozet*

1.6. Nature reserve and access areas

The southern French islands of Kerguelen, Crozet, Saint-Paul and Amsterdam were classified as a nature reserve in 2006. The management plan of the reserve defined zones of integral protection and areas of regulated access. No invertebrate species are protected, but significant steps have been taken to limit the introduction and spread of alien species. Eight zones on the Îles Kerguelen and four on Île de la Possession (Figure 1.16) are now subject to access authorization for scientific activities.

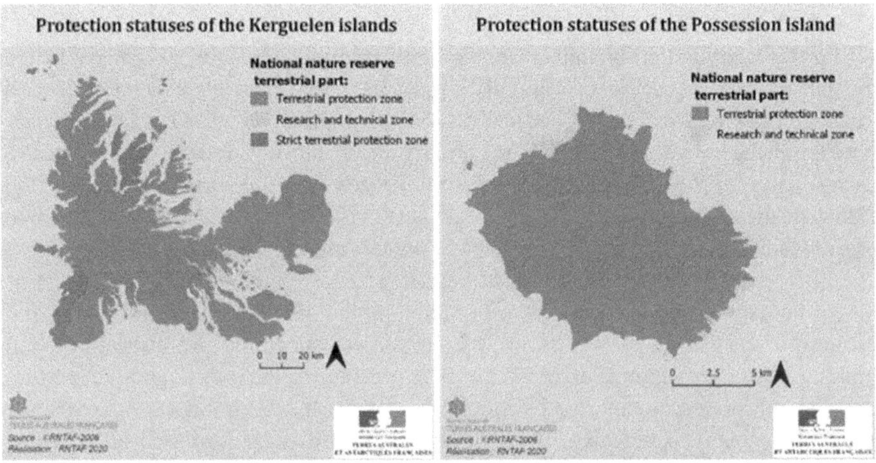

Figure 1.16. *Protected areas of the Îles Kerguelen and Île de la Possession (maps: Floran Hoarau)*

1.7. The terrestrial macroinvertebrates

Terrestrial macroinvertebrates are represented by seven classes: Clitellata, Gastropoda, Chilopoda, Arachnida, Malacostraca, Collembola and Insecta. The class of Malacostraca includes at least one species of woodlouse, and that of Chilopoda includes at least one centipede species. These two species, whose identification is uncertain, are not presented in this book. Likewise, Collembola and small invertebrates such as mites and tardigrades are not described in this book because we do not have sufficiently recent revisions. The following chapters therefore describe the earthworms, gastropods, spiders and insects currently present on both the Îles Kerguelen and Île de la Possession.

The native species described in the literature have almost all been found during the specific prospecting campaigns that have been carried out since the 2000s. This book therefore gives an account of our current state of knowledge. Similarly, the taxonomy used here takes into account the various revisions that have been published.

In regards to introduced species, only established species have been included. Species seen very occasionally, but which have not succeeded in becoming established, are not considered to be part of the fauna of the two islands. An exception is made for some occasional Lepidoptera that arrive on the islands by natural migration routes. Most of the introduced species have been identified.

For the taxonomic nomenclature, we have used the reference system published by the National Museum of Natural History in Paris[1]. The identification criteria given here are relatively brief because an identification guide and keys have already been published (Hullé *et al.* 2018). The arrows on the illustrations of the species correspond to the identification criteria written in *blue and italics* in the "identification" paragraph. For more direct access to information, we have chosen to present families and species in alphabetical order within each order.

As far as possible, we provide information on the distribution, identification and ecology of each species. The level of information concerning the distribution and biology of species depends on the knowledge we have of them through the literature and field observations reported by various scientists. It is appropriate to pay tribute here to the many people who have collected, identified and studied all of these species. It was not only specialists but also students and field assistants who contributed for a few days, weeks or months to the acquisition of this knowledge. In fact, the permanent scientific programs supported by the "Institut Polaire Français Paul-Émile Victor" (IPEV) make it possible each year to send researchers on summer campaigns and send young volunteers for a full year.

The ability of introduced species to become invasive is a major concern for scientists and managers. In this book, the currently known distribution of each introduced species is given graphically in the form of a presence map. This is the minimum information we have today on the distribution of these species because not all areas have been explored, especially on the Îles Kerguelen which are large.

1 https://inpn.mnhn.fr/programme/referentiel-taxonomique-taxref.

To complement the morphological approach, molecular characterization of native and introduced species has also been undertaken since 2016. Currently, 162 sequences (gene C01) have been deposited in the Barcode of Life Data System[2]. This is the STIMT (Sub-Antarctic Terrestrial Invertebrate Molecular Taxonomy) project, which to date includes sequences from 28 native species and 17 introduced species belonging to the different orders present on the two islands.

2 www.boldsystems.org.

2

Annelida – Clitellata

Seven species of earthworm are present on both islands: four native species of Acanthodrilidae, mainly on Île de la Possession, and three introduced species of Lumbricidae, mainly on the Îles Kerguelen. The native species all belong to the genus *Microscolex* (Frenot 1986a) (Table 2.1). All species are epigeic and feed on plant litter as well as, most probably, on organic matter provided by bird or marine mammal colonies. Due to their particularly high density, up to 2,000 ind/m^2, earthworms from sub-Antarctic islands are likely to play an extremely important role in the recycling of organic matter (Frenot 1986b).

The other Oligochaete families present (Phreodrilidae and Enchytraeidae) are not included in this book because there is no recent revision and their identification is still uncertain.

Family	Species	Îles Kerguelen	Île de la Possession
Acanthodrilidae	*Microscolex crozetensis*		Endemic
	Microscolex enzenspergeri		Endemic
	Microscolex kerguelarum	Endemic	
	Microscolex luykeni		Endemic
Lumbricidae	*Allolobophora chlorotica*	Introduced	
	Dendrodrilus rubidus	Introduced	Introduced
	Eiseniella tetraedra	Introduced	

Table 2.1. *Earthworms of the Îles Kerguelen and Île de la Possession*

2.1. Order: Crassiclitellata

2.1.1. *Family: Acanthodrilidae*

This family is represented by only one genus, *Microscolex*. A group of closely related *Microscolex* species is distributed throughout the sub-Antarctic islands of the Indian, Atlantic and Pacific oceans and the far south of the South American continent. Each of these islands has one or more species that are endemic to them (Lee 1968; Frenot 1986a).

2.1.1.1. Microscolex crozetensis *(Michaelsen, 1905)*

(Annelida: Clitellata: Crassiclitellata: Acanthodrilidae)

Distribution

Microscolex crozetensis was described by Wilhelm Michaelsen following the "*Deutsche Südpolar-Expedition*" of the *Gauss* in 1902. The species is endemic to the Îles Crozet (Tétry 1947; Frenot 1986b; Frenot *et al.* 1987).

Identification

The body is pale brown to dark brown in color, with *two pairs of lighter prostatic papillae* on segments XVII and XIX (Figure 2.1). The clitellum covers segments XIII to XVI (not visible in Figure 2.1). Four subspecies have been described, which are distinguished by the pigmentation of the body and the shape of the prostatic papillae (Bouché 1982; Joly *et al.* 1987; Blakemore 2008).

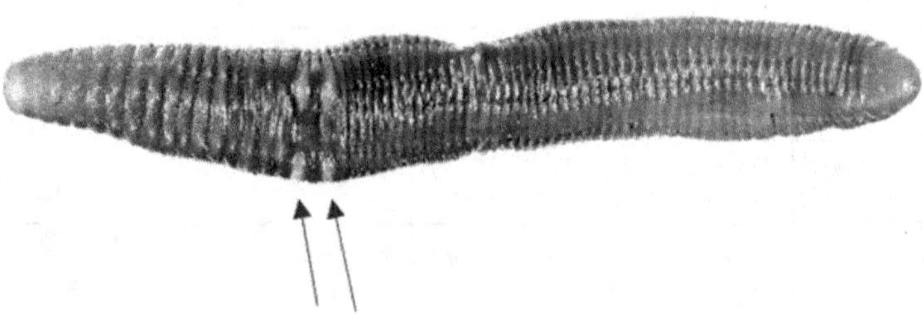

Figure 2.1. Microscolex crozetensis *(photo: B. Chaubet)*

Ecology

M. crozetensis lives in salt-enriched environments such as herbaceous coastal areas exposed to sea spray, or in more inland sites, in areas with strong air currents in halophilic vegetation (Frenot 1986b; Joly *et al.* 1987). We also sampled it at altitudes up to 500 m a.s.l. and in fellfield areas.

2.1.1.2. Microscolex enzenspergeri *(Michaelsen, 1905)*

(Annelida: Clitellata: Crassiclitellata: Acanthodrilidae)

Distribution

Microscolex enzenspergeri was also discovered during the "*Deutsche Südpolar-Expedition*" in 1902 with the *Gauss*, on board which were the scientists Josef Enzensperger and Karl Luyken. This species is endemic to the Îles Crozet (Frenot *et al.* 1987).

Identification

The color is dark brown before and just after the clitellum and otherwise light brown. The clitellum covers segments XIII to XVI. There is only *one pair of prostatic papillae*, on segment XVII (Figure 2.2). Four subspecies have been described. They are distinguished essentially by the pigmentation of their body (Frenot 1986a; Blakemore 2008).

Figure 2.2. Microscolex enzenspergeri *(photo: B. Chaubet)*

Ecology

Like the other *Microscolex*, *M. enzenspergeri* lives just below the surface of the soil. The different subspecies seem to occur in different habitats: *M. e. bicolor* in fellfield above 300 m a.s.l., *M. e. nigra* in open vegetation areas, *M. e. enzenspergeri* and *M. e. alba* mainly in organic or peat soils, as well as in low-altitude herbaceous areas and in fellfield if a vegetative cover has formed (Tréhen *et al.* 1985; Frenot 1986b).

2.1.1.3. Microscolex kerguelarum *(Grube, 1877)*

(Annelida: Clitellata: Crassiclitellata: Acanthodrilidae)

Distribution

This species was described by Adolph Eduard Grube in 1877 as *Lumbricus kerguelarum*, following its discovery by the *Challenger* expedition to the Îles Kerguelen in 1874 (Tétry 1947). It is also present on Heard Island, McDonald Island and the Prince Edward Islands (Lee 1968; Hänel 1999; Blakemore 2008; Chown and Froneman 2008; Dartnall 2017).

Identification

The body is reddish-brown, and the *clitellum* is lighter (Figure 2.3). The clitellum covers segments XIII to XVI. There are two pairs of prostatic papillae on segments XVII and XIX (not visible in Figure 2.3).

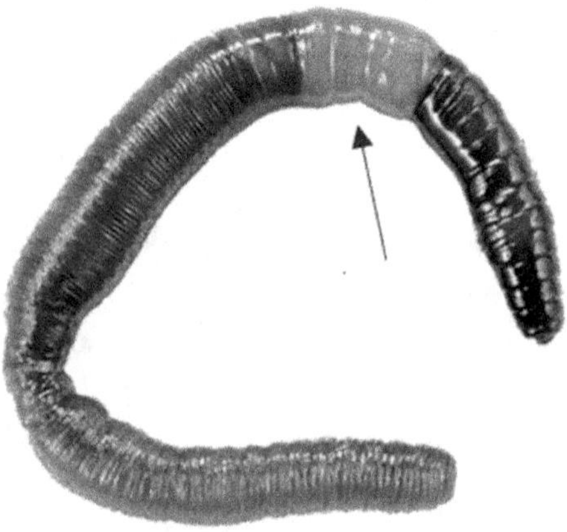

Figure 2.3. Microscolex kerguelarum *(photo: B. Chaubet)*

Ecology

The ecology of this species has been poorly described. It was caught mainly in very wet riverside or coastal soils. It is common on the Îles Kerguelen.

2.1.1.4. Microscolex luykeni *(Michaelsen, 1905)*

(Annelida: Clitellata: Crassiclitellata: Acanthodrilidae)

Distribution

Microscolex luykeni is the third earthworm species to have been described by Wilhelm Michaelsen following the *Gauss* Expedition in 1902. This species is endemic to Île de la Possession (Blakemore 2008). Its presence on the other Crozet islands is likely; however, it was not observed on Île aux Cochons during a survey conducted by Yves Frenot in 1982 (Frenot *et al.* 1987).

Identification

The body is darker in color than other *Microscolex* species. The post-clitellum segments are very dark, almost purplish, and marked with *white dots* which are very visible at the base of the chaetae. The *clitellum* covers segments XIII to XVI, and *two pairs of prostatic papillae* are present on segments XVII and XIX (Figure 2.4).

Figure 2.4. Microscolex luykeni *(photo: B. Chaubet)*

Ecology

L. luykeni lives in flooded or water-saturated environments, river beds and peat soils (Frenot 1986a). This species appears to be less frequent than the other two indigenous species *M. crozetensis* and *M. enzenspergeri.*

2.1.2. *Family: Lumbricidae*

Lumbricidae is a family of nearly 700 species. Of palearctic origin, some of them are now worldwide. Three species have been introduced to the Îles Kerguelen and Île de la Possession.

2.1.2.1. Allolobophora chlorotica *(Savigny, 1826)*

(Annelida: Clitellata: Crassiclitellata: Lumbricidae)

Distribution

Allolobophora chlorotica has been introduced to the Îles Kerguelen. This species is also mentioned in the Atlantic islands of St Helena and Tristan da Cunha

(Blakemore 2008). On the Îles Kerguelen, it has been observed only in the periphery of Port-aux-Français (Figure 2.5).

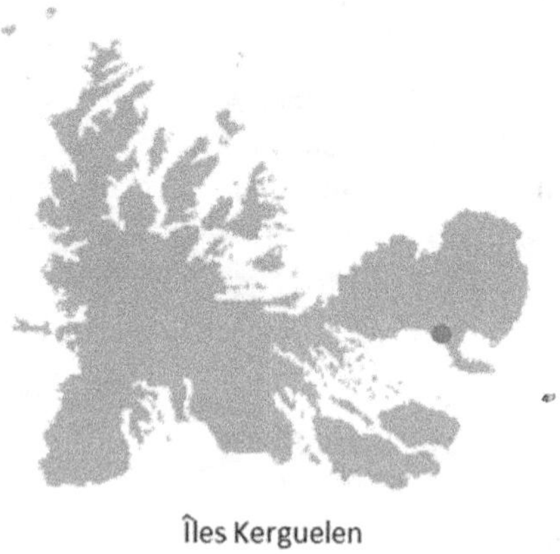

Figure 2.5. *Currently known distribution of* Allolobophora chlorotica

Identification

A. chlorotica is pale pink to greenish in color. The *clitellum* covers the dorsal surface of segments XXX to XXXVI, leaving *three pairs of papillae* visible on the ventral side (Figure 2.6). The *male pore* is very visible on segment XV (Bouché 1972).

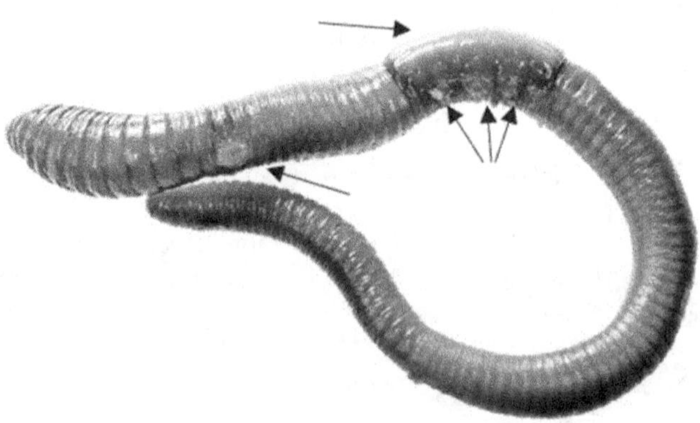

Figure 2.6. Allolobophora chlorotica *(photo: B. Chaubet)*

Ecology

This species is rare and very little is known on its ecology and habitats on the Îles Kerguelen. It has been found mainly in humid lawns bordering ponds.

2.1.2.2. Dendrodrilus rubidus *(Savigny, 1826)*

(Annelida: Clitellata: Crassiclitellata: Lumbricidae)

Distribution

Dendrodrilus rubidus is of holoarctic origin and is now found worldwide. One subspecies, *Dendrodrilus rubidus tenuis* (Eisen 1874), has been introduced to several sub-Antarctic islands in the South Indian Ocean (Heard Island, the Îles Kerguelen, the Îles Crozet, Marion Island) as well as in the South Atlantic Ocean (the Tristan da Cunha Islands group, the Falkland Islands) and the South Pacific Ocean (Campbell Island, Auckland Island, Macquarie Island) (Frenot 1985; Frenot *et al.* 2005; Greenslade 2006; Blakemore 2008; Dartnall 2017). Bouché (1982) suggested that *D. rubidus* was probably introduced to the Îles Kerguelen by sealers and whalers as early as the 19th Century. It has now become invasive on the Îles Kerguelen (Figure 2.7). On Île de la Possession, it was first observed in 1982 on Base Alfred-Faure. It is likely that it was not introduced until 1974, the date of the first introduction of potting soil for the greenhouses that were in place at that time (Frenot 1985).

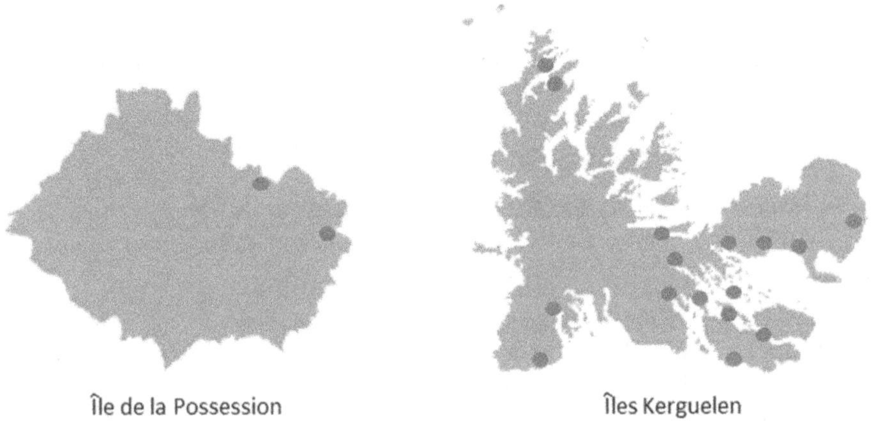

Figure 2.7. *Currently known distribution of* Dendrodrilus rubidus

Identification

D. rubidus is easily recognizable by its *pink anterodorsal coloring*. The *clitellum* covers segments XXVII to XXXI (Figure 2.8). The *male pore* is on segment XV. Three subspecies have been described, which are distinguished by the shape of the puberculum (organ located on the ventral face of the worm at clitellum level) (Frenot 1985).

Figure 2.8. Dendrodrilus rubidus *(photo: B. Chaubet)*

Ecology

D. rubidus is an epigeic earthworm, living in the uppermost level of soil, in the litter or peaty superficial horizon (Bouché 1972). On the Îles Kerguelen, it shares the same ecological niche as *Microscolex kerguelarum* but does not seem to have a negative impact on this native species (Prat *et al.* 2002). It is the most common introduced earthworm on the Îles Kerguelen.

2.1.2.3. Eiseniella tetraedra *(Savigny, 1826)*

(Annelida: Clitellata: Crassiclitellata: Lumbricidae)

Distribution

Eiseniella tetraedra is cosmopolitan and has been introduced to Macquarie Island, the Îles Kerguelen, Tristan da Cunha Islands, and St Helena (Blakemore 2008). It was not found during an intensive survey conducted on Macquarie Island in 1997–1998, suggesting that it is no longer present (Greenslade 2006). It is still scarce on the Îles Kerguelen (Figure 2.9).

Figure 2.9. *Currently known distribution of* Eiseniella tetraedra

Identification

E. tetraedra is dark in color, often reddish-brown. The *clitellum* covers segments XX to XXV (or XXI to XXVI), and the male pores are in segment XIII. Cylindrical in shape at the anterior, it becomes *quadrancular* at the posterior part (Figure 2.10).

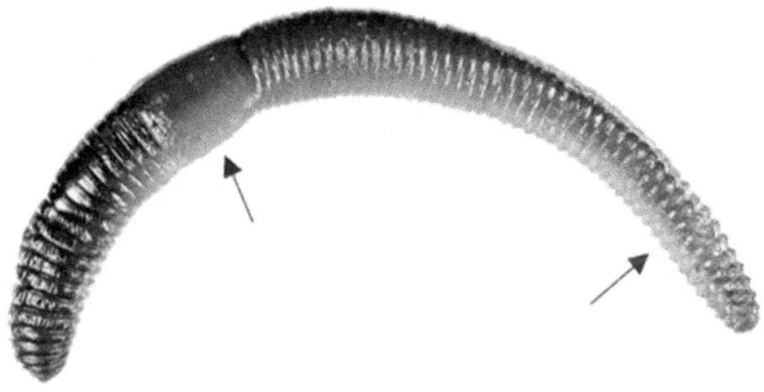

Figure 2.10. Eiseniella tetraedra *(photo: B. Chaubet)*

Ecology

E. tetraedra is generally semi-aquatic. It is often found in wet mud along the shores of lakes or rivers and in wet soil (Bouché 1972).

3

Mollusca – Gastropoda

Terrestrial gastropods are represented by three species belonging to the same order, a tiny autochtonous land snail and two introduced slugs.

3.1. Order: Stylommatophora

3.1.1. *Family: Charopidae*

3.1.1.1. Notodiscus hookeri *(Reeve, 1854)*

(Mollusca; Gastropoda; Stylommatophora; Charopidae)

Distribution

Notodiscus hookeri was discovered by Sir Joseph Dalton Hooker during the mission of the *Erebus* and the *Terror* in 1840. It is the only native land snail in the Indian Ocean sub-Antarctic islands (Heard Island, McDonald Island, the Îles Kerguelen, the Îles Crozet, the Prince Edward Islands) (Hänel 1999; Charrier *et al.* 2013) and in South Georgia (Pugh and Smith 2011).

Identification

The shell and body are brown in color. The shell is somewhat flattened. Its growth is continuous. The largest shells measure 7 to 7.5 mm (Figure 3.1).

Figure 3.1. Notodiscus hookeri *(photo: B. Chaubet)*

Ecology

N. hookeri is a litter-dwelling snail that feeds exclusively on lichens such as *Usnea taylori* and *Pseudocyphellaria crocata* (Gadea *et al.* 2017). It includes two ecophenotypes, with a mineral versus an organic shell, depending on the calcium availability. The "mineral" phenotype is common in coastal areas where mineral soils are rich in exchangeable calcium, whereas the "organic" phenotype is found preferentially in the fellfield, which is poorer in exchangeable calcium but sufficiently rich in lichen to allow the species to grow there (Charrier *et al.* 2013; Gadea *et al.* 2017).

3.1.2. *Family: Agriolimacidae*

Two species of Agriolimacidae have been identified on the Îles Kerguelen. At least one of the two species is also present on Île de la Possession, but its identification has not yet been confirmed.

3.1.2.1. Deroceras invadens *(Reise, Hutchinson, Schunack and Schlitt, 2011)*

(Mollusca; Gastropoda; Stylommatophora; Agriolimacidae)

Distribution

Deroceras invadens (previously called *D. panormitanum* and *D. caruanae*) is a cosmopolitan slug of palearctic origin (Hutchinson *et al.* 2014). The species was recorded on the Îles Kerguelen and Marion Island (*D. caruanae* in Hänel (1999) or

D. panormitanum in Lee and Chown (2016)), in the Indian Ocean Province, and on Tristan da Cunha Islands and Chatham Island (Hutchinson *et al.* 2014). Its presence on Île de la Possession is uncertain. This species is restricted to a few sites on the base of Port-aux-Français or around field huts on the Îles Kerguelen (Figure 3.2).

Figure 3.2. *Currently known distribution of* Deroceras invadens

Identification

Adults are 20–30 mm long and variable in color from light grayish-brown to dark. The skin is fairly transparent (Figure 3.3). The mucus is colorless.

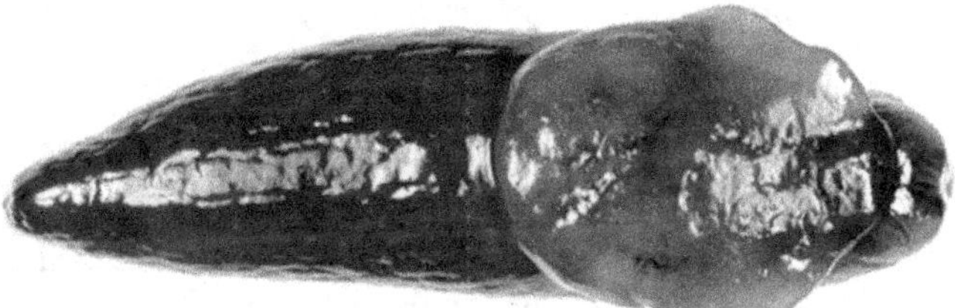

Figure 3.3. Deroceras invadens *(photo: B. Chaubet)*

Ecology

D. invadens is phytophagous and could have been introduced recently to the islands with fresh food, such as salads. On the Îles Kerguelen, it has been recorded in disturbed habitats and on grassland.

3.1.2.2. Deroceras reticulatum *(O.F. Müller, 1774)*

(Mollusca; Gastropoda; Stylommatophora; Agriolimacidae)

Distribution

Of palearctic origin, *Deroceras reticulatum* has become cosmopolitan. It has been reported on Macquarie, Campbell, Auckland and Chatham Islands (Greenslade 2006) in the New Zealand Province and on Tristan da Cunha and Gough Islands (Preece 2001; Jones *et al.* 2002) in the Atlantic Ocean. Its presence has been confirmed on the Îles Kerguelen in at least at two localities (Figure 3.4) but not yet on Île de la Possession.

Figure 3.4. *Currently known distribution of* Deroceras reticulatum

Identification

Adults can reach 40–50 mm long and are variable in color from cream gray to brown, with dark traces dorsally. The mucus is milky white in color.

Ecology

This species is phytophagous, but in the sub-Antarctic context, field observations are currently missing.

4

Arthropoda – Arachnida

Large terrestrial arachnids include spiders, harvestmen and pseudoscorpions. Mites, which belong to the microfauna, for non-parasite species, are not discussed in this chapter.

4.1. Order: Araneae

Spiders are an important functional group within ecosystems in terms of the number of species and abundance. These are the most diverse group of invertebrate predators on both islands. There are 15 species, eight native and seven introduced. Île de la Possession hosts seven native species and five introduced species. Five of the seven species native to Île de la Possession are endemic to the Archipel Crozet. In contrast, the Îles Kerguelen host five introduced species and only two native species (including one endemic) (Table 4.1).

4.1.1. *Family: Anapidae*

The family Anapidae includes approximately 230 species of minute spiders. It is represented here by a single species.

4.1.1.1. Crozetulus minutus *(Hickman, 1939)*

(Arthropoda; Arachnida; Araneae; Anapidae)

Distribution

Crozetulus minutus is currently known as a species strictly endemic to Île de la Possession, and has therefore not been observed on the other Crozet islands.

Identification

C. minutus is 1 mm long, with orange to reddish-brown prosoma and legs, and gray-brown, globular opisthosoma (Figure 4.1). The juveniles are almost white.

Figure 4.1. Crozetulus minutus *(photo: B. Chaubet)*

Family	Species	Îles Kerguelen	Île de la Possession
Anapidae	*Crozetulus minutus*		Endemic
Desidae	*Myro jeanneli*		Endemic
	Myro kerguelenensis	Sub-Antarctic	Sub-Antarctic
	Myro paucispinosus		South Indian Ocean
	Myro pumilus		Endemic
Hahniidae	*Hahnia crozetensis*		Endemic
Linyphiidae	*Neomaso antarcticus*	Endemic	
	Ringina antarctica		Endemic
	Tenuiphantes tenuis	Introduced	
	Lepthyphantes leprosus	Introduced	
	Ostearius melanopygius		Introduced
Agelenidae	*Tegeneria domestica*	Introduced	Introduced
Pholcidae	*Pholcus phalangioides*		Introduced
Theridiidae	*Steatoda grossa*	Introduced	Introduced
	Steatoda triangulosa	Introduced	Introduced

Table 4.1. *Spiders of the Îles Kerguelen and Île de la Possession*

Ecology

C. minutus lives in colonies of variable density in rather wide but confined spaces such as at the base of vegetation, mosses and under stones. It builds a horizontal geometric web of about 5 mm in diameter (Ledoux 1991). It was recorded in moorland and bog, up to 100 m in altitude (Davies 1973).

4.1.2. *Family: Desidae*

This family naturally ranges across the southern hemisphere. It is represented on the Îles Kerguelen and Île de la Possession by four *Myro* species. These species do not build traps and probably hunt by sight (Ysnel and Ledoux 1988). All four species are native and two are endemic.

4.1.2.1. Myro kerguelenensis *(O. Pickard-Cambridge, 1876)*

(Arthropoda; Arachnida; Araneae; Desidae)

Distribution

Myro kerguelenensis was one of the first native species described following the expeditions of the 19th Century (Berland 1947). *M. kerguelenensis* has been divided into two subspecies, *M. k. kerguelenensis* (O. Pickard-Cambridge, 1876) on the Îles Kerguelen, Heard and McDonald Islands and *M. k. crozetensis* (Enderlein, 1909) on the Îles Crozet and Prince Edward Islands (Gressitt 1970; Ledoux 1991; Pugh 2004). The species is also present on the Macquarie and Bishop Islands (Gressitt 1970; Davies *et al.* 1997; Greenslade 2006). The presence of *M. kerguelenensis* on Macquarie Island could result from ballooning (Pugh 2004).

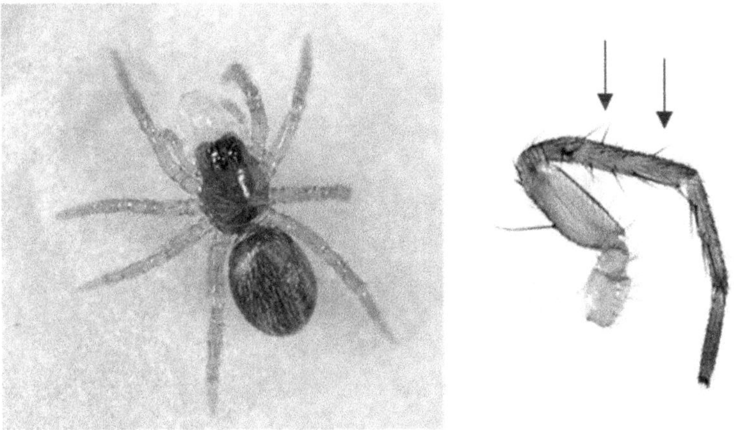

Figure 4.2. Myro kerguelenensis*: juvenile eating a mite and foreleg tibia dorsal spines (photos: B. Chaubet)*

Identification

The prosoma is 1.9–2.6 mm long and the body 6–7 mm for females and 4–5 mm for males. The foreleg tibiae have *two dorsal spines*. The opisthosoma is clear with many white spots, especially in *M. k. crozetensis*. The prosoma has slightly darker radiant marks (Figure 4.2).

Ecology

M. kerguelenensis is the most common spider on both islands. It can easily be observed roaming on the soil in search of prey. It is a generalist predator of insects, isopods, earthworms and even the snail *Notodiscus hookeri*. Adults are present all year round, with hatching in spring and autumn. The species is present in all environments, from the seaside to high altitude. It is particularly abundant where there is a concentration of organic matter (penguin rookeries, seaside under plant cover, proximity to petrel burrows, etc.) (Ysnel and Ledoux 1988).

4.1.2.2. Myro jeanneli *(Berland, 1947)*

(Arthropoda; Arachnida; Araneae; Desidae)

Distribution

Myro jeanneli was discovered during the *Bougainville* cruise in 1939 (Berland 1947). This species is endemic to the Îles Crozet. However, it has not been recorded on Île des Pingouins (Gressitt 1970; Ledoux 1991).

Figure 4.3. Myro jeanneli, *female (left) and male (right) (photos: B. Chaubet)*

Identification

M. jeanneli is a quite large spider. The prosoma is 2.8 mm long in males and 4–5 mm long in females. The total body length is 9–10 mm long for females and 5–7 mm for males, therefore indicating a significant sexual dimorphism. The body is rather dark: the prosoma is gray with almost black radiating bands; the opisthosoma is black with light spots and light chevrons. There are no dorsal spines on foreleg tibiae (Figure 4.3).

Ecology

M. jeanneli is characteristic of the fellfield where it is common (Davies 1973; Ledoux 1991). It is also found at the upper limit of vegetation in *Agrostis magellanica* areas. This spider hunts by day and likes to frequent ground with stones that serve as a refuge. It is active all year round (Ledoux 1991).

4.1.2.3. Myro paucispinosus *(Berland, 1947)*

(Arthropoda; Arachnida; Araneae; Desidae)

Distribution

Myro paucispinosus was also discovered during the *Bougainville* expedition in 1939 (Berland 1947). This native species is present in the Archipel Crozet (Île aux Pingouins, Île de l'Est and Île de la Possession) and on the Prince Edward Islands (Gressitt 1970; Ledoux 1991; Pugh 2004).

Figure 4.4. Myro paucispinosus*: female, and foreleg (photos: B. Chaubet)*

Identification

M. paucispinosus is brown in color, with lighter radiating lines on the prosoma and clear geometric patterns on the opisthosoma. The prosoma of females is 3.5–5 mm long, and that of males is 2.5–4 mm. The spider is covered with dense hairs. There are *very few spines on the foreleg tibiae but there are long trichobothria* (Figure 4.4).

Ecology

M. paucispinosus is uncommon on Île de la Possession, where it is limited to areas of dense vegetation. It seems to avoid the fellfield where *M. jeanneli* is abundant (Ledoux 1991).

4.1.2.4. Myro pumilus *(Ledoux, 1991)*

(Arthropoda; Arachnida; Araneae; Desidae)

Distribution

The discovery of *Myro pumilus* by Lewis Davies dates back to the 1970s. This is one of the last native species described. It is endemic to the Îles Crozet where, currently, it has only been observed on the eastern group of islands (Île de la Possession and Île de l'Est) (Ledoux 1991).

Identification

M. pumilus can be easily distinguished from other *Myro* species by its very small size (body 2.5 mm long), its solid pale gray color, *the eye arrangement, which constitutes a very dark mass* and very large posterior eyes, occupying the entire width of the prosoma (Figure 4.5).

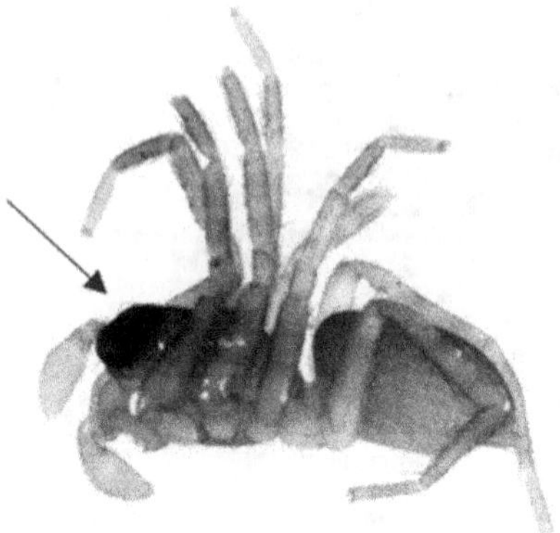

Figure 4.5. Myro pumilus, *male (photo: B. Chaubet)*

Ecology

M. pumilus can go unnoticed because of its small size. It is, however, abundant in the fellfield and under the fronds of *Blechnum penna-maritima*. It is found in the narrow spaces between gravels and plants where it hunts by sight (Ledoux 1991).

4.1.3. *Family: Hahniidae*

Hahniidae is a fairly homogeneous family, spread over all continents. They are small spiders that weave webs at ground level. This family is represented here by only one species.

4.1.3.1. Hahnia crozetensis *(Hickman 1939)*

(Arthropoda; Arachnida; Araneae; Hahniidae)

Distribution

Hahnia crozetensis is endemic to the Îles Crozet (Île de la Possession, Île de l'Est and Île aux Cochons) (Dreux 1970; Ledoux 1991). It was discovered in 1939 during the *Bougainville* expedition (Berland 1947).

Identification

H. crozetensis is a spider of variable size (3–4 mm long on average), light brown in color. The prosoma and legs are pale brown. The opisthosoma is grayish-white, with long dorsal hairs. The *spinnerets* are very visible and form a line (exclusive feature of this family). The two exteriors are particularly long and clearly extend beyond the opisthosoma in dorsal view (Figure 4.6).

Figure 4.6. Hahnia crozetensis *(photo: B. Chaubet)*

Ecology

H. crozetensis is abundant everywhere, in narrow spaces where humidity remains high (under stones half sunk into the ground, under mosses, under dense vegetation). It weaves an irregular, tightly woven web and stands underneath. It is rarely caught in traps (Ledoux 1991).

4.1.4. *Family: Linyphiidae*

Linyphiidae is the most diversified family of spiders, including more than 4,300 species worldwide. They are commonly known as sheet weavers from the shape of their webs that some species extend over the vegetation. Here they include two endemic and three introduced species.

4.1.4.1. Neomaso antarcticus *(Hickman, 1939)*

(Arthropoda; Arachnida; Araneae; Linyphiidae)

Distribution

Neomaso antarcticus is endemic to the Îles Kerguelen (Berland 1947; Ledoux 1991). It is one of the two species of spider native to this island.

Identification

N. antarcticus is a small spider (2–3 mm long). Its color is uniform from honey yellow to dark brown. The legs are brown, thin and relatively short (Figure 4.7).

Figure 4.7. Neomaso antarcticus *(photo: B. Chaubet)*

Ecology

N. antarcticus is abundant, especially in meadows dominated by *Azorella selago*. It weaves a web in very confined spaces and sometimes deep underground.

4.1.4.2. Ringina antarctica *(Hickman, 1939)*

(Arthropoda; Arachnida; Araneae; Linyphiidae)

Distribution

Ringina antarctica is probably endemic to Île de la Possession, but its presence has yet to be sought on the other islands of the Archipel Crozet (Ledoux 1991).

Identification

R. antarctica is a small spider (about 2 mm). It is grayish-green with a yellow tinge. The opisthosoma is dorsally marked with three chevrons and two lighter stripes.

Ecology

This species is found by carefully exploring confined spaces in vegetation holes or under large stones, but its population densities remain relatively low. It is a sedentary spider that builds a small horizontal web under which it rests (Ledoux 1991).

4.1.4.3. Tenuiphantes tenuis *(Blackwall, 1852)*

(Arthropoda; Arachnida; Araneae; Linyphiidae)

Distribution

Tenuiphantes tenuis is a palearctic species that has been introduced to north Atlantic islands, New Zealand, North and South America (Vink and Kean 2013) and, more recently, to Antipodes Island (Pugh 2004) and the Îles Kerguelen. On the Îles Kerguelen, it has been recorded only in the southeast sector, the sector most frequented by humans (Figure 4.8).

Figure 4.8. *Currently known distribution of* Tenuiphantes tenuis

Identification

T. tenuis is 2.5–3 mm long. The prosoma is dark brown and the opisthosoma marbled with strong contrasts, the ratio of dark and light parts being very variable. The legs are elongated and light brown to orange (Figure 4.9).

Figure 4.9. Tenuiphantes tenuis, *female (photo: B. Chaubet)*

Ecology

T. tenuis can be locally abundant, especially in the vicinity of human installations. This spider weaves webs at the top of plants. Web networks can cover very large areas and contain several hundred individuals (Figure 4.10).

Figure 4.10. *Grasses covered by the webs of a* Tenuiphantes tenuis *colony at Île Longue (Kerguelen) (photo: M. Plantegenest)*

4.1.4.4. Lepthyphantes leprosus *(Ohlert, 1865)*

(Arthropoda; Arachnida; Araneae; Linyphiidae)

Distribution

Lepthyphantes leprosus is present in the northern hemisphere (North America, Europe and Russia) and has been introduced at various sites in the southern hemisphere. Recently introduced to the Îles Kerguelen, it has remained confined to Port-aux-Français (Figure 4.11). The species has also been reported from Gough Island (Hänel 2007).

Figure 4.11. *Currently known distribution of* Lepthyphantes leprosus

Identification

L. leprosus is 2.5–4 mm long. The opisthosoma has gray-brown chevrons on a light background. The legs are pale orange and may have slightly marked rings on the femurs (Figure 4.12). It can be confused with *T. tenuis*, and only an examination of the genital organs (epigyne and pedipalps) can distinguish them.

Figure 4.12. Lepthyphantes leprosus, *female (photo: B. Chaubet)*

Ecology

L. leprosus has only been observed in some buildings of Port-aux-Français.

4.1.4.5. Ostearius melanopygius *(O. Pickard-Cambridge, 1879)*

(Arthropoda; Arachnida; Araneae; Linyphiidae)

Distribution

This species is cosmopolitan by introduction. Originally from South America, it has been introduced across Europe, Turkey, North Africa, South Africa, China and New Zealand. It was first observed on Île de la Possession in 2012 (Rapp 2013) where it is currently confined to the Base Alfred-Faure (Figure 4.13). It has also been reported from Île Saint-Paul in the Indian Ocean (Berland 1947).

Figure 4.13. *Currently known distribution of* Ostearius melanopygius

Identification

O. melanopygius is 2–3 mm long. The body has a characteristic coloration with a brown-green prosoma marked with dark radiant streaks and an orange opisthosoma with a *black tip*, giving the Latin name of the species (Figure 4.14).

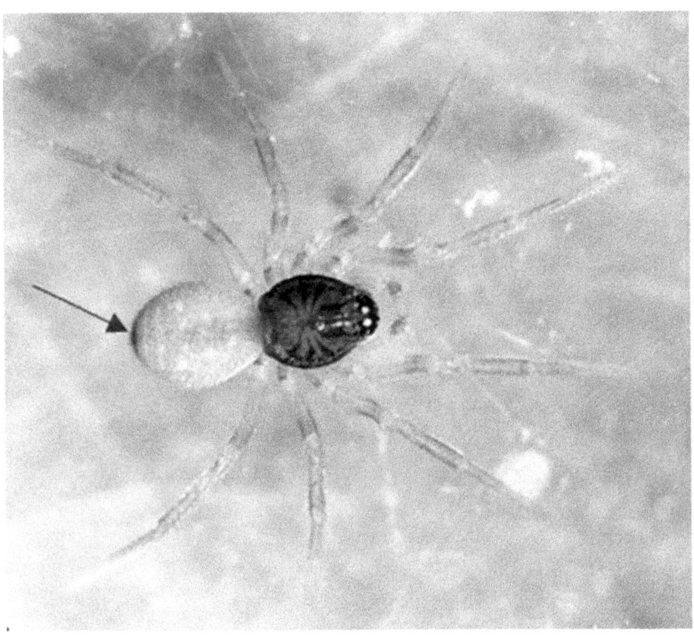

Figure 4.14. Ostearius melanopygius *(photo: A. Pierre and M. Bertrand)*

Ecology

This species remains in or in the immediate vicinity of the buildings of the Base Alfred-Faure.

4.1.5. *Family: Agelenidae*

The family Agelinidae is represented by only one species, introduced to the Îles Kerguelen.

4.1.5.1. Tegeneria domestica *(Clerk, 1758)*

(Arthropoda; Arachnida; Araneae; Agelinidae)

Distribution

Tegenaria domestica is a cosmopolitan species that was recorded for the first time on the Îles Kerguelen in the 2000s (Frenot *et al.* 2005) (Figure 4.15). In 2012, it was actively sought on Île de la Possession but was never found (Rapp 2013). The species has also been recorded from South Georgia (Pugh 2004) and Gough Island (Hänel 2007).

Figure 4.15. *Currently known distribution of* Tegenaria domestica

Identification

T. domestica is a medium-sized spider, 6–10 mm long. The body is light beige with a darker prosoma. It has strong hairs, especially on the opisthosoma and legs (Figure 4.16).

Figure 4.16. Tegenaria domestica *(photo: B. Chaubet)*

Ecology

Confined to Port-aux-Français, this species frequents the dark parts of buildings, such as basements. It weaves a funnel web where it usually waits for its prey. Many cuticles of the introduced beetle, *Merizodus soledadinus*, are easily observed on or at the foot of these spider webs.

4.1.6. *Family: Pholcidae*

This family is cosmopolitan. It is represented on the islands by only one introduced species.

4.1.6.1. Pholcus phalangioides *(Fuesslin, 1775)*

(Arthropoda; Arachnida; Araneae; Pholcidae)

Distribution

Originally restricted to warmer parts of the west Palearctic, *Pholcus phalangioides* is now cosmopolitan. Introduced to Île de la Possession, it is currently restricted to the buildings of the Base Alfred-Faure (Figure 4.17).

Figure 4.17. *Currently known distribution of* Pholcus phalangioides

Identification

Very long legs that make it look like harvestmen (Opiliones) characterize this species (Figure 4.18). Its gray-brown body has dark markings (often three spots on each side). The prosoma also has a darker spot in its center.

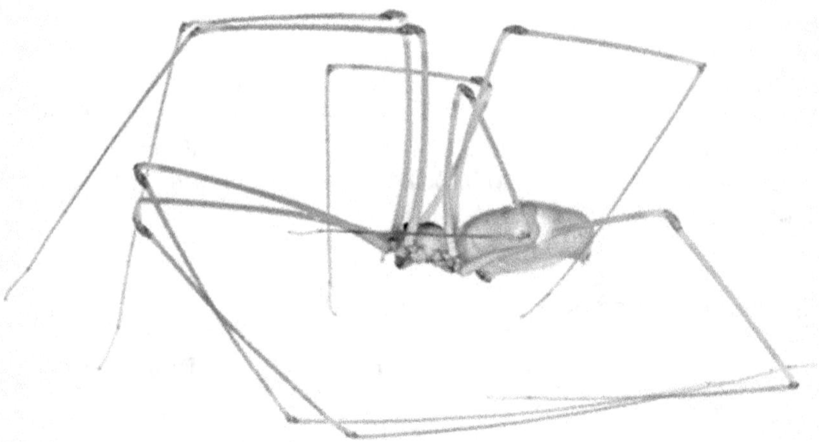

Figure 4.18. Pholcus phalangioides *(photo: B. Chaubet)*

Ecology

P. phalangioides is a typical house spider. It can be observed on the ceilings of rooms where it weaves an irregular web. It clings to it upside down, its opisthosoma pointing upwards.

4.1.7. *Family: Theridiidae*

Theridiidae is a highly diversified spider family, distributed worldwide. On the Îles Kerguelen and Île de la Possession, it is represented by two introduced species of the genus *Steatoda* that share the same distribution and have very similar ecology.

Distribution of both *Steatoda* species

Frenot *et al.* (2005) report only one Theridiidae species, *Steatoda triangulosa*, present in the buildings of Port-aux-Français (Îles Kerguelen). Rapp (2013) reported this same species from Base Alfred-Faure (Île de la Possession) for the first time in 2012. We have since found a second species, *S. grossa*, in the buildings of both bases, which suggests a very recent introduction (Figure 4.19).

Ecology of both *Steatoda* species

Both species live in the buildings and construct irregular tangles of sticky fibers.

Figure 4.19. *Currently known distribution of* Steatoda triangulosa *and* S. grossa

4.1.7.1. Steatoda grossa *(C.L. Koch, 1838)*

(Arthropoda; Arachnida; Araneae; Theridiidae)

Identification

Steatoda grossa is 6–8 mm long. It has a highly contrasted opisthosoma in juveniles of both genera and adult males, but the patterns vary widely from one individual to another. The adult female's opisthosoma is very low in contrast or even uniformly light brown. The prosoma and legs are orange-brown in color (Figure 4.20).

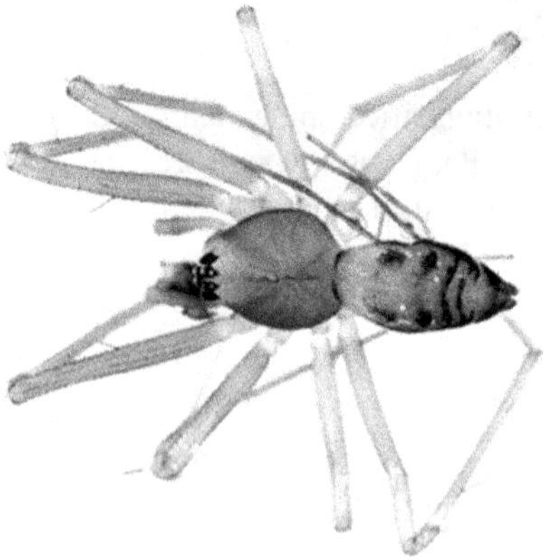

Figure 4.20. Steatoda grossa, *male (photo: B. Chaubet)*

4.1.7.2. Steatoda triangulosa *(Walckenaer, 1802)*

(Arthropoda; Arachnida; Araneae; Theridiidae)

Identification

Steatoda triangulosa is 4–6 mm long. Its coloration with a high-contrast opisthosoma and an orange prosoma is characteristic (Figure 4.21).

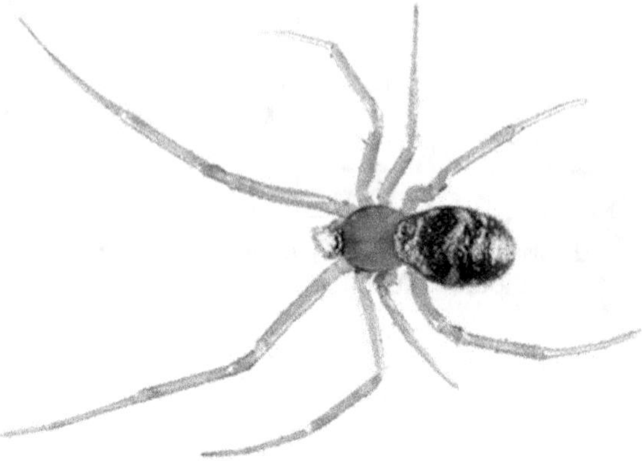

Figure 4.21. Steatoda triangulosa *(photo: B. Chaubet)*

4.2. Order: Opiliones

A few Opiliones species are reported from sub-Antarctic islands, such as Campbell Island, Auckland Island, Snares Island and the Falkland Islands (Beron 2018).

4.2.1. *Family: Triaenonychidae*

4.2.1.1. Nuncia unifalculata *(Enderlein, 1909)*

(Arthropoda; Arachnida; Opiliones; Triaenonychidae)

Distribution

Nuncia unifalculata is endemic to the Archipel Crozet. It has been reported from Île de la Possession and Île aux Cochons. It was described by Gunther Enderlein after the *Gauss* expedition of 1902 (Fage 1940; Dreux 1970).

Identification

N. unifalculata is 5–5.5 mm long. Prosoma and opisthosoma are brown and are not separated by a visible constriction. The long legs, with their numerous segments, are gray. The *eyes are grouped* on a central mound in front of the body. The front legs carry *strong hooks on the trochanter*. The chelicerae are very strong and adorned with teeth (Figure 4.22).

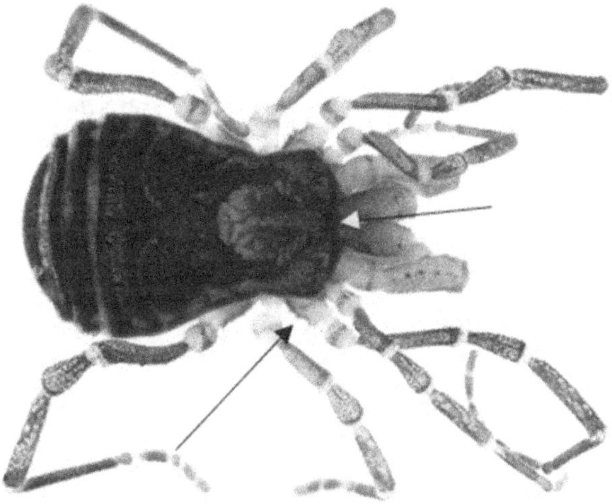

Figure 4.22. Nuncia unifalculata *(photo: B. Chaubet)*

Ecology

The biology of the species is not well known. It is probably omnivorous, like most harvestmen. On Île de la Possession, it has been observed at many sites, ranging from vegetative areas to xeric fellfield. Generally solitary, it is sometimes found in groups of about 10 individuals before overwintering.

4.3. Order: Pseudoscorpiones

4.3.1. *Family: Chthoniidae*

The family Chthoniidae contains 30 genera, including the genus *Austrochthonius* Chamberlin, which has a southern distribution from South America to Australia and New Zealand (Vitali di Castri 1968).

4.3.1.1. Austrochthonius insularis *(Vitali di Castri, 1968)*

(Arthropoda; Arachnida; Pseudoscorpiones; Chthoniidae)

Distribution

Austrochthonius insularis is only known on the Île de la Possession where it was first recorded by Jean-François Voisin in 1966 (Dreux 1970).

Identification

A. insularis is 1.5–2 mm long and pale orange in color. The cephalothorax is almost square and united. The abdomen is segmented. Two pairs of single eyes are present in the lateral position. The *pedipalps* are transformed into long, scorpion-like pincers and frame the two chelicerae, which are themselves also in the shape of pincers (Figure 4.23).

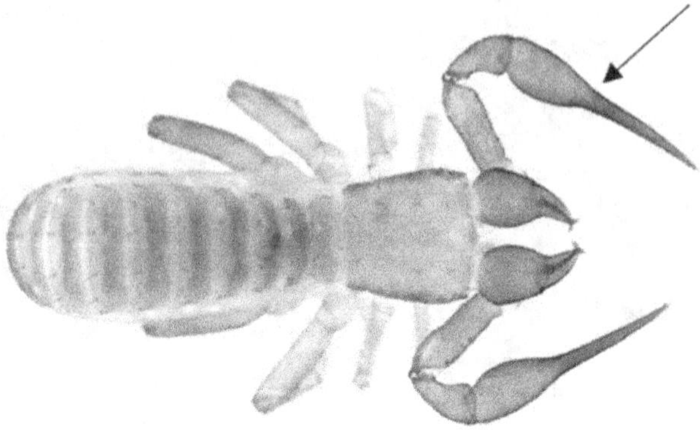

Figure 4.23. Austrochthonius insularis *(photo: B. Chaubet)*

Ecology

A. insularis is widespread throughout Île de la Possession and leads a discreet life beneath stones in well-drained areas. It is an active micro-predator that feeds upon small mites and springtails. Medium-altitude rock falls seem to be the preferred habitat and is absent in the coastal zone. It is a solitary species in general, but some gathering of individuals can sometimes be observed (Rapp 2013).

5

Arthropoda – Insecta

Insects represent 83% and 86% of the free-living macro-arthropods present on Île de la Possession and the Îles Kerguelen, respectively. Parasitic insects (Siphonaptera and Phthiraptera) are not presented here because the species have not been recently revised. Free-living insects are represented by seven orders of which Coleoptera and Diptera are the most diversified. There is currently a total of 77 insect species, including 54 indigenous and 23 non-indigenous (introduced and established). 25 species are common to both islands. Île de la Possession houses 42 indigenous species, of which 29 are strict endemics and the Îles Kerguelen houses 22 species, of which only six are strict endemics (Table 5.1).

	Îles Kerguelen		**Île de la Possession**	
Order	**Indigenous**	**Non-indigenous**	**Indigenous**	**Non-indigenous**
Coleoptera	12	4	22	1
Diptera	6	10	14	9
Hemiptera	0	6	1	5
Hymenoptera	1	0	1	0
Lepidoptera	2	0	3	0
Psocoptera	1	1	1	1
Thysanoptera	0	1	0	1
Total	**22**	**22**	**42**	**17**

Table 5.1. *Number of indigenous and non-indigenous insect species*

Among the foreign and now established insects, 12 species can be considered invasive: a beetle (*Merizodus soledadinus*), eight flies (*Fucellia tergina, Calliphora vicina, Hemeromyia (?)* sp., *Limnophyes minimus, Lycoriella sativae, Psychoda parthenogenetica*, *Smittia* sp., *Trichocera maculipennis*), two aphids (*Myzus ascalonicus*, *Rhopalosiphum padi*) and one thrips (*Apterothrips apteris*).

5.1. Order: Coleoptera

With 32 native species and only four introduced, Coleoptera is the most characteristic order of the entomofauna of the South Indian Ocean Province. 24 species are endemic to either Îles Crozet or Îles Kerguelen, seven species are also present on other sub-Antarctic islands of the South Indian Ocean Province and one in different sub-Antarctic provinces (Table 5.2).

Family	Species	Îles Kerguelen	Île de la Possession
Curculionidae	*Bothrometopus angusticollis*	Endemic	
	Bothrometopus brevis	South Indian Ocean	
	Bothrometopus comes		Endemic
	Bothrometopus crozetensis		Endemic
	Bothrometopus daviesi		Endemic
	Bothrometopus desolationis		Endemic
	Bothrometopus fasciatus		Endemic
	Bothrometopus gracilipes	South Indian Ocean	
	Bothrometopus gravis		Endemic
	Bothrometopus randi		South Indian Ocean
	Bothrometopus sulcatus[1]	Endemic	
	Canonopsis sericea	South Indian Ocean	
	Christensenia antarctica		Endemic
	Disker tenuicornis	Endemic	
	Ectemnorhinus bougainvillei		Endemic
	Ectemnorhinus drygalskii	Endemic	
	Ectemnorhinus fuscus	Endemic	
	Ectemnorhinus geniculatus		Endemic

1 *B. sulcatus* is only known from one specimen captured by R. Jeannel in 1939 in western Kerguelen. It is so similar to *B. fasciatus* and *B. comes* from Crozet, that Jean-François Voisin suspects a labeling error due to poor working conditions on the ship at the time (Voisin *et al.* 2016, 2017).

Family	Species	Îles Kerguelen	Île de la Possession
	Ectemnorhinus richtersi		Endemic
	Ectemnorhinus viridis	South Indian Ocean	
	Neocanonopsis dreuxi		Endemic
	Palirhoeus eatoni	South Indian Ocean	South Indian Ocean
	Xanium possessionense		Endemic
	Xanium vanhoeffenianum		Endemic
Hydraenidae	*Meropathus chuni*	End	
	Meropathus randi		South Indian Ocean
Latridiidae	*Cartodere nodifer*	Introduced	Introduced
	Latridius minutus	Introduced	
Ptinidae	*Ptinus tectus*	Introduced	
Staphylinidae	*Antarctotachinus crozetensis*		Endemic
	Leptusa atriceps	Sub-Antarctic	Sub-Antarctic
	Pseudoplectus antarcticus		Endemic
Trechidae	*Amblystogenium minimum*		Endemic
	Amblystogenium pacificum		Endemic
	Merizodus soledadinus	Introduced	
	Temnostega antarctica		Endemic

Table 5.2. *Coleoptera of the Îles Kerguelen and Île de la Possession*

5.1.1. *Family: Curculionidae*

Weevils are the most diverse beetle family and the most characteristic of the sub-Antarctic fauna. The 24 species present are indeed native, 18 of which are endemic to one or the other island. For weevil names, we followed the nomenclature of Kuschel and Chown (1995).

5.1.1.1. Bothrometopus angusticollis *(C.O. Waterhouse, 1875)*

(Arthropoda; Insecta; Coleoptera; Curculionidae)

Distribution

Bothrometopus angusticollis is endemic to the Îles Kerguelen. It is one of the species discovered during the *Challenger* expedition in 1874.

Identification

B. angusticollis is quite large (6–12 mm), generally elongated with well-defined *square shoulders*. The scales are fine, numerous and copper brown in color. Some individuals may have a clearer drawing on the elytra (Figure 5.1).

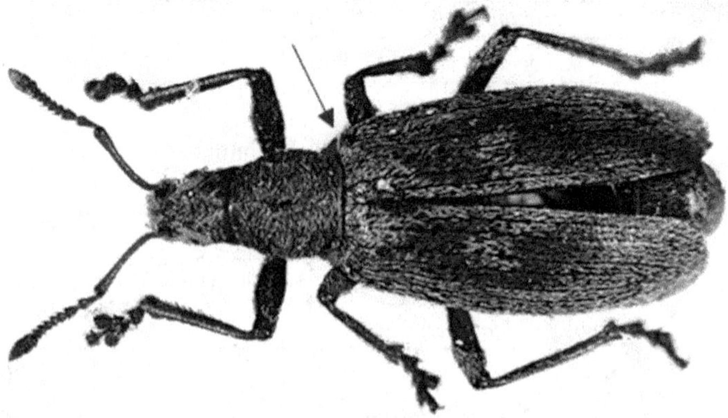

Figure 5.1. Bothrometopus angusticollis *(Collection MNHN[2], photo: H. Baird)*

Ecology

B. angusticollis lives in coastal vegetation and feeds on cryptogamous plants. It finds refuge under rocks (Chown 1989). Previously cited as abundant, it has become rare, perhaps due to the decrease in *Pringlea antiscorbutica*, the Kerguelen cabbage (Jeannel 1940). This species has only been observed four times around Golfe du Morbihan in the past 10 years.

5.1.1.2. Bothrometopus brevis *(C.O. Waterhouse, 1875)*

(Arthropoda; Insecta; Coleoptera; Curculionidae)

Distribution

Bothrometopus brevis is present on the Îles Kerguelen as well as on Heard and McDonald Islands (Kuschel 1970; Chown and Convey 2016).

---

2 Muséum National d'Histoire Naturelle, Paris, France.

Identification

B. brevis is a small weevil, 3–5 mm long, with a stocky silhouette, well pronounced shoulders and a shiny appearance. The green scales are irregularly distributed, giving the elytra a marbled appearance (Figure 5.2).

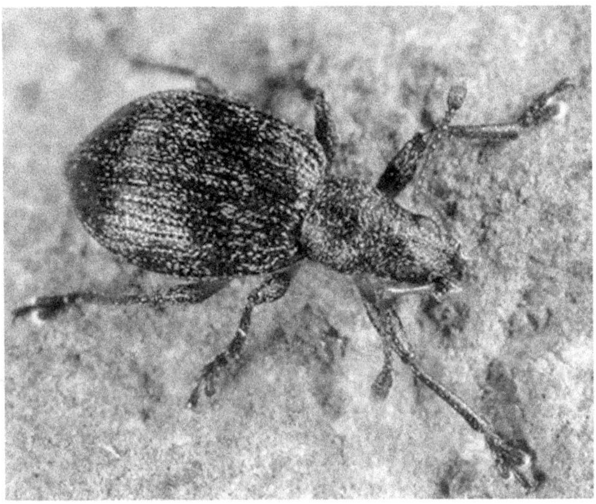

Figure 5.2. Bothrometopus brevis
(photo: B. Chaubet)

Ecology

B. brevis is very common on coastal rock faces in the supralittoral zone (Voisin *et al.* 2017). The species can also be found inland in fellfield. Adults are mainly active on warm and humid days; otherwise, they can take refuge in rock crevices (Chown *et al.* 2004). Adults and larvae feed on algae, cyanobacteria and lichen and occasionally on bryophytes (Chown and Klok 2001b).

5.1.1.3. Bothrometopus comes *(Chown and Kuschel, 1994)*

(Arthropoda; Insecta; Coleoptera; Curculionidae)

Distribution

Bothrometopus comes is known from Île de la Possession and Île de l'Est (a group of eastern islands in the Archipel Crozet) (Chown and Kuschel 1994).

Identification

B. comes is 5.5–8 mm long. The body is black, the legs are black-brown and the antennae are brown-red. The *scutellum* is small and densely pubescent with light brown scales. Elytra are elongated, rounded and shiny with a well-developed pubescence (Figure 5.3). This species resembles *B. fasciatus* and is distinguished by the shape of the pronotum in the lateral view: convex in *B. comes* and flat in *B. fasciatus*.

Figure 5.3. Bothrometopus comes *(photo: B. Chaubet)*

Ecology

B. comes is one of the last native species described. Its biology is poorly known. It seems to depend on the fellfield or rocky or stony areas (Voisin *et al.* 2017).

5.1.1.4. Bothrometopus crozetensis *(Enderlein, 1904)*

(Arthropoda; Insecta; Coleoptera; Curculionidae)

Distribution

Bothrometopus crozetensis is endemic to the Îles Crozet. It has been reported on all the islands of this archipelago (Chown 1994).

Identification

B. crozetensis is 3–4 mm long. It is very similar to *B. brevis* (Dreux and Voisin 1986a) and is distinguished by its elongated shape and more granular integument of the pronotum (Figure 5.4).

Figure 5.4. Bothrometopus crozetensis *(Collection MNHN, photo: H. Baird)*

Ecology

B. crozetensis lives mainly along the coast. Like *B. brevis*, it is found on rocks near the coast and on lawns. It can go into the fellfield up to 150 m a.s.l. It is active during the day, especially in mild weather. It feeds on unicellular algae or mosses on rocks. In bad weather, it hides in cracks or under mosses (Voisin *et al.* 2017).

5.1.1.5. Bothrometopus daviesi *(Chown and Kuschel, 1994)*

(Arthropoda; Insecta; Coleoptera; Curculionidae)

Distribution

Bothrometopus daviesi is endemic to Île de la Possession. The species was discovered by Lewis Davies in 1968 (Chown and Kuschel 1994).

Identification

B. daviesi is a small weevil (2–3 mm long). It is rather similar to *B. gracilipes* but much smaller (Chown and Kuschel 1994). The body is dark-brown and covered with small, almost invisible hairs; the pronotum is smooth (Figure 5.5).

Figure 5.5. Bothrometopus daviesi *(Collection MNHN, photo: M. Hullé)*

Ecology

We know almost nothing about the ecology of this species. We have only captured it four times in the past 10 years and always in the xeric fellfield.

5.1.1.6. Bothrometopus desolationis *(Jeannel, 1940)*

(Arthropoda; Insecta; Coleoptera; Curculionidae)

Distribution

Bothrometopus desolationis is endemic to Île de la Possession. Its presence at Île de l'Est (Chown 1994) has yet to be confirmed (Voisin *et al.* 2017).

Identification

B. desolationis is one of the smallest sub-Antarctic weevils (2–3 mm long). Black in color, the body is covered with many fine, gray-brown scales, thin and widened at the tip. The legs and antennae are reddish-brown. The *pronotum* is more long than wide (Figure 5.6).

Figure 5.6. Bothrometopus desolationis *(photo: B. Chaubet)*

Ecology

B. desolationis is discreet and relatively unobserved due to its small size, coloration and immobility reflex (Voisin *et al.* 2017). Voisin *et al.* (2017) describe it as a fellfield species, but we have also found it in coastal areas and along rivers. *B. desolationis* is an epilithic species that feeds on cryptogams (Chown 1989).

5.1.1.7. Bothrometopus fasciatus *(Jeannel, 1940)*

(Arthropoda; Insecta; Coleoptera; Curculionidae)

Distribution

B. fasciatus is *a priori* endemic to Île de la Possession and Île de l'Est (Dreux 1971b). It was discovered by René Jeannel during the *Bougainville* cruise in 1939 (Jeannel 1940).

Identification

B. fasciatus is 6–9 mm long. The pronotum is granular. The metallic green scales do not completely cover the elytra, drawing poorly delineated figures. The species is close to *B. comes* and is mainly distinguished by the shape of the pronotum, which is almost flat in the lateral view in *B. fasciatus* and convex in *B. comes* (Figure 5.7).

Figure 5.7. Bothrometopus fasciatus *(MNHN, photo: H. Baird)*

Ecology

B. fasciatus is not common. Curiously, Dreux (1965b) considered it very common when he first visited Île de la Possession in 1962. This species can be found under stones *a priori* in the fellfield up to 700 m a.s.l. It probably feeds on mosses (Voisin 1976).

5.1.1.8. Bothrometopus gracilipes *(C.O. Waterhouse, 1875)*

(Arthropoda; Insecta; Coleoptera; Curculionidae)

Distribution

Bothrometopus gracilipes is present on the Îles Kerguelen and Heard Island, where it seems less frequent (Jeannel 1940; Kuschel 1970; Dreux 1978).

Identification

B. gracilipes is a medium-sized weevil (4–5 mm). It is dark in color and not very shiny. It is covered with dense hairs of lighter color and evenly distributed over the pronotum and elytra (Figure 5.8).

Figure 5.8. Bothrometopus gracilipes
(photo: B. Chaubet)

Ecology

B. gracilipes is common and sometimes abundant. It is found in various environments from the coastal zone to the fellfield. Jeannel (1940) described it as frequent beneath the stones of the rocky mounds of the sunny slopes of *Acaena magellanica*. It feeds on algae, cyanobacteria, lichens and occasionally on bryophytes (Chown 1989; Chown and Klok 2001b).

5.1.1.9. Bothrometopus gravis *(Chown and Kuschel, 1994)*

(Arthropoda; Insecta; Coleoptera; Curculionidae)

Distribution

Bothrometopus gravis is endemic to Île de la Possession. It was recently described from specimens collected by Lewis Davies in 1962 (Chown and Kuschel 1994).

Identification

B. gravis is 7–9 mm long. The pronotum is smooth. *Elytra are dull* and striae have shallow punctures (Figure 5.9). *B. gravis* is similar to *B. randi* whose elytra are brighter and striae deeper. Voisin *et al.* (2017) consider them to be two forms of the same species, *B. randi*.

Figure 5.9. Bothrometopus gravis
(photo: B. Chaubet)

Ecology

B. gravis lives in stony areas under stones. It can be found up to 600–700 m a.s.l. (Chown and Kuschel 1994).

5.1.1.10. Bothrometopus randi *(Jeannel, 1953)*

(Arthropoda; Insecta; Coleoptera; Curculionidae)

Distribution

Bothrometopus randi is present on the Îles Crozet and on the Prince Edward Islands (Voisin 1976; Dreux and Voisin 1984, 1992; Chown 1992, 1994; Chown and Convey 2016).

Identification

B. randi is 5.5–9 mm long. Elytra are shiny. Antennae, legs and mandibles have rusty reflections (Figure 5.10).

Figure 5.10. Bothrometopus randi
(photo: B. Chaubet)

Ecology

B. randi is an epilithic species that lives mainly on rock walls and rock tops from the coastal zone to 750 m a.s.l. (Chown 1992; Voisin *et al.* 2017). On Marion Island, it is also found in fellfield (Hänel 1999). It feeds on algae, lichens and mosses (Chown 1992, 1993; Chown and van Drimmelen 1992).

5.1.1.11. Canonopsis sericea *(C.O. Waterhouse, 1875)*

(Arthropoda; Insecta; Coleoptera; Curculionidae)

Distribution

Canonopsis sericea is endemic to the Îles Kerguelen, Heard and McDonald Islands (Jeannel 1940; Kuschel 1970; Chown and Convey 2016).

Identification

C. sericea is large in size (12–16 mm), elongated and gray-brown in color. Its body is covered with a dense and fluffy pubescence. The *pronotum* has four tubers giving a bumpy appearance. The *femurs* are swollen (Figure 5.11).

Figure 5.11. Canonopsis sericea
(photo: B. Chaubet)

Ecology

C. sericea is *a priori* confined to the southeast of the Îles Kerguelen, particularly on the islands of Golfe du Morbihan. It seems that this species, linked to the presence of Kerguelen cabbage, *Pringlea antiscorbutica*, had been common before, but has become rare with the decline of Kerguelen cabbage in favor of *Acaena magellanica* (Jeannel 1940; Voisin *et al.* 2017). We found it only twice during the surveys of the past 10 years. On Heard Island, *C. sericea* is found in all vegetation types, excluding the supralittoral zone and rock faces devoid of mosses. Adults feed on vascular plants including *Poa cookii*, *Azorella selago* and *Acaena magellanica* and on bryophytes (Kuschel and Chown 1995; Chown and Klok 2001b).

5.1.1.12. Christensenia antarctica *(Brinck, 1945)*

(Arthropoda; Insecta; Coleoptera; Curculionidae)

Distribution

Christensenia antarctica is endemic to the Îles Crozet. The species is present on all the islands of the archipelago: Îles aux Pingouins, Îlots des Apôtres, Île aux

Cochons, Île de la Possession and Île de l'Est. It was discovered during a Norwegian expedition in 1908 (Dreux 1971b; Dreux and Voisin 1983).

Identification

C. antarctica is a large weevil (10–18 mm) of a dark brown color, sometimes a little yellowish or greenish. The elytra have convex and *bumpy interstriae* and have long erected silks. The *pronotum* is flat in the lateral view (Figure 5.12).

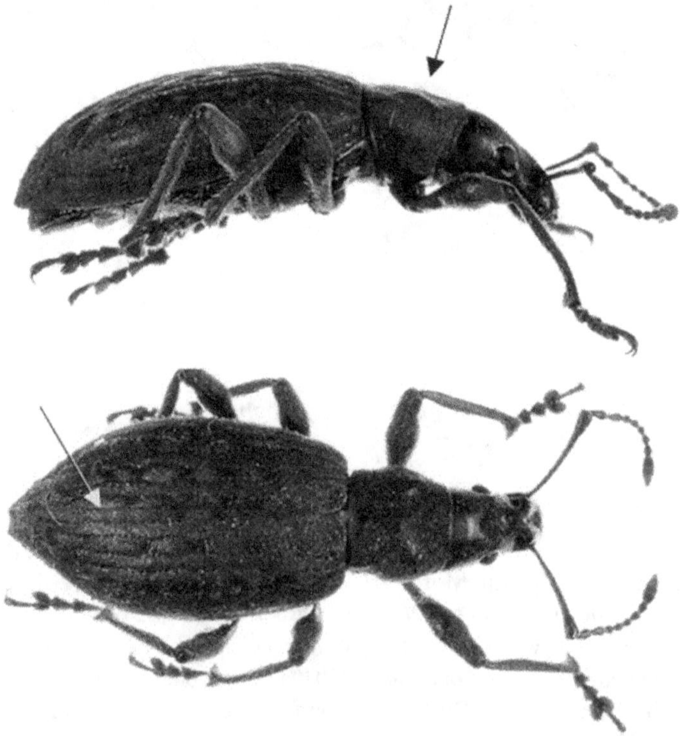

Figure 5.12. Christensenia antarctica *(photo: B. Chaubet)*

Ecology

Fellfield and stony environments are the typical habitats of *C. antarctica* (Vernon and Voisin 1990; Bailly and Dreux 1992; Dreux and Voisin 1992). The species lives under rocks where it can be abundant. It seems to be absent from coastal vegetated areas. It is therefore an epilithic species that feeds on angiosperms and bryophytes (Chown 1989; Kuschel and Chown 1995).

5.1.1.13. Disker tenuicornis *(Jeannel, 1940)*

(Arthropoda; Insecta; Coleoptera; Curculionidae)

Distribution

Disker tenuicornis is endemic to the Îles Kerguelen. It was discovered by René Jeannel during the *Bougainville* cruise in 1939, that stopped in Baie des Swains (southeast coast) (Jeannel 1940; Dreux and Voisin 1987).

Identification

D. tenuicornis is a small weevil, 2–4 mm long and dark brown in color. Elytra have long, slightly dense hairs and gray-green scales. Striae are strongly punctuated. The tibiae have a *mucron* on their distal part, and the *claw segment* is long (Figure 5.13).

Figure 5.13. Disker tenuicornis *with details of mucron and claw segment (photo: B. Chaubet)*

Ecology

This species has rarely been observed: in 1939 by René Jeannel in Baie des Swains, then in 1992 by Philippe Dreux in the same area and at Port-Couvreux, then we only found it again in 2018 in Baie des Swains and nearby (Presqu'Ile Jeanne-d'Arc). Because of its stony habitat, Kuschel and Chown (1995) consider it to feed on algae, cyanobacteria and lichens.

5.1.1.14. Ectemnorhinus bougainvillei *(Jeannel 1940)*

(Arthropoda; Insecta; Coleoptera; Curculionidae)

Distribution

Ectemnorhinus bougainvillei is endemic to Île de la Possession and Île de l'Est (Îles Crozet eastern group) (Dreux and Voisin 1986b).

Identification

E. bougainvillei is 6–9 mm long. The body is dark and shiny. It is covered by golden green scales. Antennae and legs are lighter, from yellowish to reddish-brown (Figure 5.14).

Figure 5.14. Ectemnorhinus bougainvillei *(MNHN, photo: H. Baird)*

Ecology

E. bougainvillei lives in lowland vegetation where it feeds on angiosperms (Chown 1989, 1992). On Île de l'Est, it can also be found in fellfield with *E. geniculatus* (Voisin *et al.* 2017).

5.1.1.15. Ectemnorhinus drygalskii *(Enderlein, 1909)*

(Arthropoda; Insecta; Coleoptera; Curculionidae)

Distribution

Ectemnorhinus drygalskii is endemic to the Îles Kerguelen. It was discovered during the *Gauss* expedition in 1902 (Jeannel 1940).

Identification

E. drygalskii is 5–9 mm long. Scales are bluish-green; they are few in number and geometrically arranged on the elytra (Figure 5.15).

Figure 5.15. Ectemnorhinus drygalskii *(photo: B. Chaubet)*

Ecology

The species has been reported from several sites on the Îles Kerguelen but it is apparently uncommon (Voisin *et al.* 2017). We only observed it twice in the vegetation during our field research. According to Chown (1989), it feeds on vascular plants.

5.1.1.16. Ectemnorhinus fuscus *(Enderlein, 1903)*

(Arthropoda; Insecta; Coleoptera; Curculionidae)

Distribution

Ectemnorhinus fuscus is endemic to the Îles Kerguelen. It was discovered during the *Gauss* expedition in 1902, by Emil Werth (Voisin *et al.* 2017).

Identification

E. fuscus is 5–8 mm long. The scales are green or yellow and always have bronze reflections. *E. fuscus* has often been confused with *E. viridis*, which is morphologically close, although it is larger and thinner. Synonymized by Dreux and Voisin (1988), Kuschell and Chown (1995) have given it back its status as a valid species.

Ecology

We have never found this species in all of our investigations. Chown (1989) considers that it would feed on vascular plants.

5.1.1.17. Ectemnorhinus geniculatus *(Jeannel, 1940)*

(Arthropoda; Insecta; Coleoptera; Curculionidae)

Distribution

Ectemnorhinus geniculatus is endemic to Île de la Possession and Île de l'Est (Dreux 1971b; Dreux and Voisin 1986b). However, Voisin *et al.* (2017) consider that its presence on Île de l'Est must be confirmed. Dreux (1965a) also reported it from Île aux Cochons in the western group of the Îles Crozet.

Identification

E. geniculatus is a large weevil, 6–11 mm long with long legs. Elytra are dark and shiny. They have short, green-golden scales. The *tarsal pads* are elongated (Figure 5.16).

Figure 5.16. Ectemnorhinus geniculatus *and tarse details (MNHN, photo: H. Baird)*

Ecology

D. geniculatus lives in wetlands at low elevations. It is found under stones or between the roots of plants. It can also be found near colonies of penguins. It feeds on angiosperms or plant debris (Dreux 1965b; Chown 1992; Voisin *et al.* 2017).

5.1.1.18. Ectemnorhinus richtersi *(Enderlein, 1904)*

(Arthropoda; Insecta; Coleoptera; Curculionidae)

Distribution

Ectemnorhinus richtersi is endemic to the Îles Crozet, where it occurs on all the islands except Île aux Pingouins, where it is replaced by *E. pluricro* (Dreux 1965a, 1971b; Dreux and Voisin 1978, 1986b; Kuschel and Chown 1995). These two species are reclassified as subspecies by Voisin *et al.* (2017). It was discovered in 1902, during the *Deutsche Südpolar-Expedition* (Dreux 1965b).

Identification

E. richtersi is 5–9 mm long. The integument is dark, bright and covered with dense scales, yellow-green to bluish-green. The scales are evenly distributed over the interstriae except on certain parts where the integument is bare, which gives dark lines. The *legs are orange-brown* with the *front tibiae distinctly incurved* at the tip (Figure 5.17).

Figure 5.17. Ectemnorhinus richtersi *(photo: B. Chaubet)*

Ecology

E. richtersi was discovered on Île de la Possession by Ernst Vanhöffen and Ferdinand Richters under stones in mosses (Jeannel 1940). It can be found in herbaceous vegetation or in fellfield, where it is rarer. It is polyphagous, feeding on native and introduced grasses like the Kerguelen cabbage, *Pringlea antiscorbutic*a, and *Azorella selago* (Dreux and Voisin 1978; Chown 1992; Voisin *et al.* 2017).

5.1.1.19. Ectemnorhinus viridis *(G.R. Waterhouse, 1853)*

(Arthropoda; Insecta; Coleoptera; Curculionidae)

Distribution

Ectemnorhinus viridis is present on the Îles Kerguelen and Heard Island (Kuschel and Chown 1995; Chown and Convey 2016).

Identification

E. viridis is 5.5–9 mm long. It shows considerable variation in size and color and can be red-brown, brown, yellow, green or bluish. This variability explains why it has been described under different species or subspecies names (see Voisin *et al.* (2017)). The shoulders are square. The elytra are shiny and the scales are fine (Figures 5.18 and 5.19).

Figure 5.18. Ectemnorhinus viridis *(photo: B. Chaubet)*

Figure 5.19. *Color variation in* Ectemnorhinus viridis *(photo: B. Chaubet)*

Ecology

E. viridis is the most common weevil species of the Îles Kerguelen. Specimens of different colors may frequent the same sites and mate. *E. viridis* is described as eurytopic, inhabiting from lowland vegetation to fellfield. It is also found in dry or humid places such as river banks. It is very common to find it under the stones. It is polyphagous, feeding on the foliage or roots of different species of vascular plants and bryophytes (Chown 1989, 1992; Chown and Klok 2001b; Chown *et al.* 2004; Voisin *et al.* 2017).

5.1.1.20. Neocanonopsis dreuxi *(Hoffmann, 1964)*

(Arthropoda; Insecta; Coleoptera; Curculionidae)

Distribution

Neocanonopsis dreuxi is endemic to the Îles Crozet: it is present on Île de la Possession, Île de l'Est and Île des Pingouins (Dreux 1966; Dreux and Voisin 1983, 1993). The likely absence of this species from Île aux Cochons is not yet understood. Voisin *et al.* (2017) hypothesize that predation by domestic mice (*Mus musculus*) may have eliminated them.

Identification

N. dreuxi is a large, dark brown weevil (9–12 mm). Elytra have *bumpy interstriae*. Erected silks are as short as scales. The *pronotum* is convex in the lateral view (Figure 5.20).

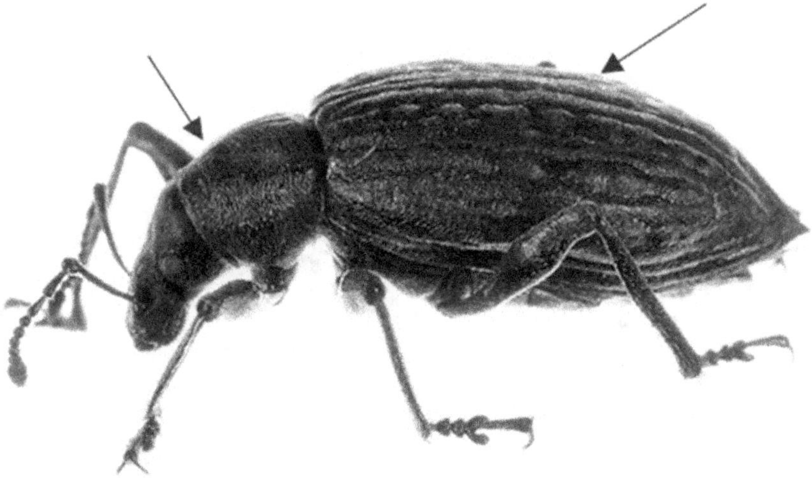

Figure 5.20. Neocanonopsis dreuxi *(photo: B. Chaubet)*

Ecology

N. dreuxi is abundant on Île de l'Est and Île des Pingouins. On Île de la Possession, it could be limited by predation from black rats (*Rattus rattus*) (Voisin *et al.* 2017). This species lives mainly in the fellfield up to 400 m a.s.l., under stones but can also descend to sea level (Dreux 1965b; Chown 1992; Dreux and Voisin 1992). It feeds on angiosperms and bryophytes (Chown 1989, 1992).

5.1.1.21. Palirhoeus eatoni *(C.O. Waterhouse, 1876)*

(Arthropoda; Insecta; Coleoptera; Curculionidae)

Distribution

Palirhoeus eatoni is endemic to the South Indian Ocean Province and is therefore the only weevil species present on all of the islands (Jeannel 1940, 1953a; Dreux 1971b; Crafford *et al.* 1986; Dreux and Voisin 1992; Chown and Klok 2001a; Chown and Convey 2016). The species was discovered in 1874 by Alfred Edwin Eaton during the *Volage* expedition, organized for observing the transit of Venus (Jeannel 1940).

Identification

P. eatoni is 3–6 mm long. It is black in color with the *top of the femurs a reddish-black*. Its silhouette is stocky with widened elytra, showing a *flatness* particularly visible in females. Elytra scales are green and scattered, leaving black areas. The *claw segments* are very long (Figure 5.21).

Figure 5.21. Palirhoeus eatoni *(photo: B. Chaubet)*

Ecology

P. eatoni lives exclusively in rocky shore and splash zones exposed to sea sprays. It is oligophagous, feeding only on a few marine algae species like Enteromorphs. It can be considered as a submarine insect which is rare in weevils. Adults and larvae are active on wet days on algae and lichen; otherwise, they shelter in rock crevices (Figure 5.22) (Davies 1973; Crafford *et al.* 1986; Chown and van Drimmelen 1992; Chown 1993, 1994; Mercer *et al.* 2000; Chown and Klok 2001b).

Figure 5.22. Palirhoeus eatoni *entering a crevice in reverse (photo: B. Chaubet)*

5.1.1.22. Xanium possessionense *(Hoffmann, 1966)*

(Arthropoda; Insecta; Coleoptera; Curculionidae)

Distribution

Xanium possessionense is endemic to Île de la Possession (Chown and Kuschel 1994). Voisin *et al.* (2015) reported its presence in the west and north of the island. We made the same observations in our recent surveys.

Identification

X. possessionense is 4–6.5 mm long. It has green-brown or blackish scales all over its body as well as erect setae, thickened at the tip (Figures 5.23 and 5.24).

Figure 5.23. Xanium possessionense *(photo: B. Chaubet)*

Figure 5.24. *Erect setae of* Xanium possessionense *and* X. vanhoeffenianum *(drawing: F. Duval)*

Ecology

The biology of this species is very poorly known. It is likely to be close to that of *X. vanhoeffenianum* which feeds on cryptogams (Voisin *et al.* 2017).

5.1.1.23. Xanium vanhoeffenianum *(Enderlein, 1904)*

(Arthropoda; Insecta; Coleoptera; Curculionidae)

Distribution

Xanium vanhoeffenianum is present on Île de la Possession and Île de l'Est (Dreux and Voisin 1985). It was discovered during the *Gauss* expedition in 1902 (Dreux 1966).

Identification

X. vanhoeffenianum is covered with large erect setae strongly thickened at the tip. Scales are brown or blackish and can be densely imbricate (Figures 5.24 and 5.25).

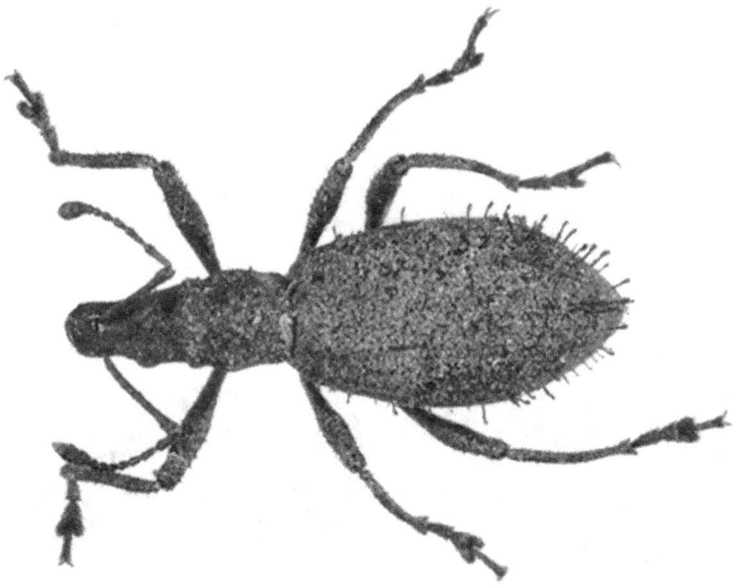

Figure 5.25. Xanium vanhoeffenianum
(photo: B. Chaubet)

Ecology

X. vanhoeffenianum lives under the stones in the fellfield and on the surface of coastal rocks (Davies 1973; Chown 1989). It can also be found on grass roots and mosses (Voisin *et al.* 2017). It feeds on bryophytes, algae, cyanobacteria and lichens (Chown 1989).

5.1.2. *Family: Hydraenidae*

Hydraenidae is a family of small aquatic beetles. Only the genus *Meropathus* occurs on the Îles Kerguelen and Île de la Possession. This genus comprises 12 species with a southern circumpolar distribution (Delgado and Palma 1998).

5.1.2.1. Meropathus chuni *(Enderlein, 1901)*

(Arthropoda; Insecta; Coleoptera; Hydraenidae)

Distribution

Meropathus chuni is present only on the Îles Kerguelen and Heard Island (Gressitt 1970).

Identification

Meropathus chuni is a small beetle (2–3 mm long) with the three parts of the body (head, thorax and abdomen) well separated. The elytra are strongly punctuated. Its color is dark with lighter areas. Hairiness is low. The male has two *hornlike processes* on the head (Figure 5.26).

Figure 5.26. Meropathus chuni *male (left) and female (right) (photos: B. Chaubet)*

Ecology

M. chuni lives along the coast, just above the seashore, in the wet mud oozing from the rocks, almost embedded in the ground under the pebbles. It is also found in *Grimmia amblyophylla* mosses and freshwater puddles and marsh (Jeannel 1940, 1953a; Voisin *et al.* 2017).

5.1.2.2. Meropathus randi *(Jeannel, 1953)*

(Arthropoda; Insecta; Coleoptera; Hydraenidae)

Distribution

Meropathus randi is present on the Îles Crozet (Îles aux Cochons, Île des Pingouins, Île de la Possession and Île de l'Est) and on the Prince Edward Islands (Marion Island and Prince Edward Island) (Jeannel 1953a; Dreux 1966; Dreux and Voisin 1969; Gressitt 1970; Chown and Convey 2016).

Identification

M. randi looks like *M. chuni*. It is small (2–3 mm) and black with tan reflections. It is mainly distinguished by the presence of curved yellow setae forming longitudinal bands on the elytra.

Ecology

M. randi has a coastal distribution (Dreux 1966; Davies 1973), but it can also be found inland up to an altitude of 60 m a.s.l. on oozing rocks, or in caves and shelters under rocks (Dreux and Voisin 1992; Voisin *et al.* 2017). It feeds on microscopic algae (Crafford *et al.* 1986).

5.1.3. *Family: Latridiidae*

Latridiidae is a cosmopolitan family of small beetles that mainly consume molds or even decomposing organic matter. Two species have been introduced to the Îles Kerguelen and Île de la Possession.

5.1.3.1. Cartodere nodifer *(Westwood, 1839)*

(Arthropoda; Insecta; Coleoptera; Latridiidae)

Distribution

Cartodere nodifer was first recorded in 1985 on Île de la Possession near the Base Alfred-Faure (Vincent and Voisin 1991) and then in 2007 in the same area (Voisin *et al.* 2017). On the Îles Kerguelen, it was first observed in 2007 at Port-aux-Français. Since then, it seems that it has begun to colonize new territories on the Îles Kerguelen but not on Île de la Possession (Figure 5.27). This species has also been reported from the Falkland Islands (Chown and Convey 2016).

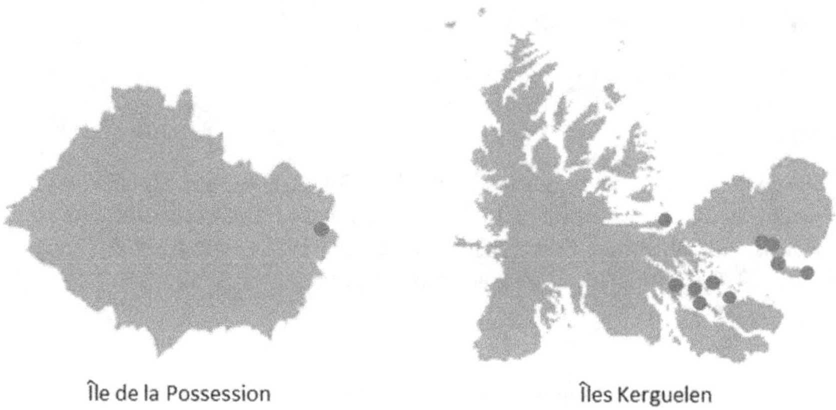

Figure 5.27. *Currently known distribution of* Cartodere nodifer

Identification

C. nodifer is a small beetle less than 2 mm long. It is black in color. Its elytra are strongly punctuated and have a *bumpy appearance*. The lateral edges of the pronotum are translucent and may have a *gelatinous appearance* (Figure 5.28). It is common to observe young adults not yet melanized that present a golden yellow coloration.

Figure 5.28. Cartodere nodifer *(photo: B. Chaubet)*

Ecology

C. nodifer is mycetophagous. It feeds on mycelia, especially molds, found on decaying plants in wetlands (Bouget and Vincent 2008).

5.1.3.2. Latridius minutus *(Linnaeus, 1767)*

(Arthropoda; Insecta; Coleoptera; Latridiidae)

Distribution

Latridius minutus was observed once in 2007 in one of the greenhouses of Base Alfred-Faure on Île de la Possession, but not again since. It was observed for the second time in the wild on the Îles Kerguelen (Île Mayes, Île Australia and Île aux Cochons in Golfe du Morbihan) during a sampling campaign in 2017–2018 (Figure 5.29). The species has also been reported from Campbell Island in the New Zealand Province (Chown and Convey 2016).

Figure 5.29. *Currently known distribution of* Latridius minutus

Identification

L. minutus is a small beetle 2 mm long. It is ochre brown in color. The elytra have *strongly punctured striae*. The pronotum is extended forward by a *protrusion* on each side of the head (Figure 5.30).

Figure 5.30. Latridius minutus *(photo: B. Chaubet)*

Ecology

L. minutus has been caught on the ground in dense grasslands dominated by native plants such as *Acaena magellanica* (Isabelle Badenhausser, pers. comm.). It probably feeds on the litter of these plants.

5.1.4. *Family: Ptinidae*

Ptinidae are small, round-bodied, flightless beetles with long and slender legs. One species has been introduced to both the Îles Kerguelen and Île de la Possession.

5.1.4.1. Ptinus tectus *(Boieldieu, 1856)*

(Arthropoda; Insecta; Coleoptera; Ptinidae)

Distribution

Ptinus tectus has been introduced to the Îles Kerguelen and Île de la Possession, where it is confined to the Base Alfred-Faure. On the Îles Kerguelen, the first record is said to date from 1963 at Port-aux-Français (Bellés *et al.* 1989). Curiously, we did not observe it in the buildings of the station but only in the huts of some islands in Golfe du Morbihan (Figure 5.31). The species has also been introduced to Campbell Island, Macquarie Island, Auckland Island and Snares Island in the New Zealand Province (Wise 1964; Watt 1971; Marris 2000; Greenslade 2006) and to South Georgia and Gough Island in the South Atlantic Ocean Province (Vogel and Nicolai 1983; Gaston *et al.* 2002).

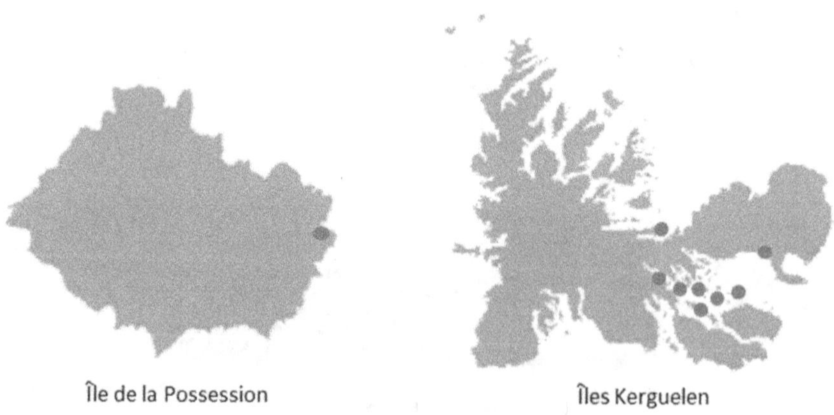

Figure 5.31. *Currently known distribution of* Ptinus tectus

Identification

P. tectus is 2–4 mm long. It is tan-brown in color, its silhouette is round and its *head* is as if encapsulated under the *thorax*. It is highly pubescent, which gives it a fluffy appearance (Figure 5.32).

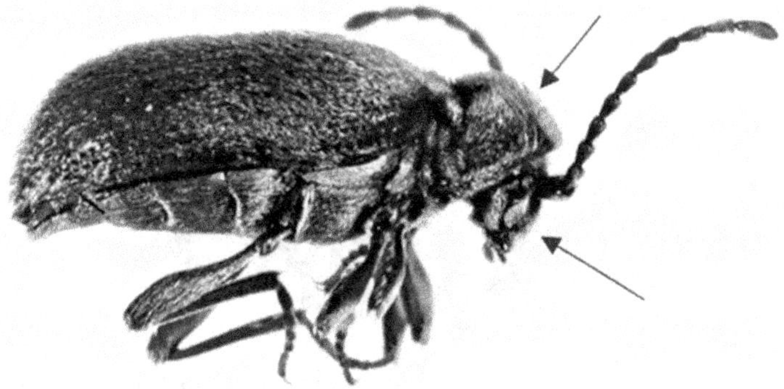

Figure 5.32. Ptinus tectus *(photo: B. Chaubet)*

Ecology

It is likely that *P. tectus* is a synanthropic species that feeds on stored starchy foods (Bellés *et al.* 1989).

5.1.5. *Family: Staphylinidae*

Staphylinidae is a family of beetles primarily distinguished by their short elytra. Most species are predators of insects and other invertebrates. Three native species are recorded from both the Îles Kerguelen and Île de la Possession.

5.1.5.1. Antarctotachinus crozetensis *(Enderlein, 1909)*

(Arthropoda; Insecta; Coleoptera; Staphylinidae)

Distribution

Antarctotachinus crozetensis is endemic to the Îles Crozet, being present on all the islands (Dreux 1965a; Vernon and Voisin 1990; Dreux and Voisin 1992; Voisin *et al.* 2017). It was discovered during the *Deutsche Südpolar-Expedition* in 1902 (Dreux 1965b).

Identification

A. crozetensis is 4–6 mm long. It is bright black in color and elongated oval in shape (Figure 5.33).

Figure 5.33. Antarctotachinus crozetensis *(photo: B. Chaubet)*

Ecology

A. crozetensis lives in the vegetation of lawns, moors and bogs. It is also found in or near bird nests such as petrels or albatrosses (Dreux and Voisin 1969; Davies 1973; Joly *et al.* 1987; Vernon and Voisin 1990; Voisin *et al.* 2017).

5.1.5.2. Leptusa atriceps *(C.O. Waterhouse, 1875)*

(Arthropoda; Insecta; Coleoptera; Staphylinidae)

Distribution

Leptusa atriceps was discovered by Alfred Edwin Eaton during the expeditions to observe the transit of Venus in 1874. This sub-Antarctic species occurs in the Archipel Crozet (Île aux Cochons, Île de la Possession and Île de l'Est), the Îles Kerguelen, Prince Edward Islands (Marion Island and Prince Edward Island), South Georgia and the Falkland Islands (Jeannel 1953a; Dreux 1966; Gressitt 1970; Steel 1970; Crafford *et al.* 1986; Orlov *et al.* 2019). It is the only Staphylinidae species of the Îles Kerguelen.

Identification

L. atriceps is small (3.5–5 mm) and thin. Its body has contrasting colors: the head and the first four visible abdominal segments are dark brown, while the thorax, elytra and tip of the abdomen are light brown. The tegument is pubescent (Figure 5.34). The insect bends easily under stress.

Figure 5.34. Leptusa atriceps *(photo: B. Chaubet)*

Ecology

Currently, *L. atriceps* seems to be more common on the Îles Kerguelen than on Île de la Possession. It is often found on pebble beaches as well as in inland vegetation (lawn, tussock, etc.). It feeds on decomposing plant materials such as decayed algae, or litter from plants enriched with organic matter near bird nests. This species can also be predator and feeds on microinvertebrates (Steel 1970; Crafford *et al.* 1986; Hänel 1999; Frank and Ahn 2011).

5.1.5.3. Pseudeuplectus antarcticus *(Enderlein, 1909)*

(Arthropoda; Insecta; Coleoptera; Staphylinidae)

Distribution

Pseudeuplectus antarcticus was discovered during the *Gauss* expedition by Ferdinand Richters. It is endemic to Île de la Possession and Île aux Cochons. Its presence is to be sought on the other islands of the archipelago (Jeannel 1940, 1953b; Voisin 1986).

Identification

P. antarcticus is very small (1.5–2 mm). It is brown to light brown in color. The head is triangular with *two depressions* on top. The body is punctuated and hairy. The legs are short and strong (Figure 5.35).

Figure 5.35. Pseudoplectus antarcticus *(photo: B. Chaubet)*

Ecology

Voisin *et al.* (2017) describe it as abundant on Île de la Possession. We did not find it very often during our surveys. At low elevations (<80 m a.s.l.), *P. antarcticus* lives in grasslands with *Acaena magellanica* or *Pringlea antiscorbutica* and near bird nests such as albatrosses and giant petrels. Above, it can be found in the *Azorella selago* cushions. It avoids the coastal meadows dominated by *Crassula moschata* and *Leptinella plumosa* (Dreux 1966; Jarrige and Voisin 1968; Voisin *et al.* 2017).

5.1.6. *Family: Trechidae*

Trechidae is a family close to Carabidae. They are represented by three native species on the Îles Crozet and one introduced species on the Îles Kerguelen. They are all predators. The genus *Amblystogenium*, described by Enderlein (1905), is endemic to the Îles Crozet.

5.1.6.1. Amblystogenium minimum *(Luff, 1972)*

(Arthropoda; Insecta; Coleoptera; Trechidae)

Distribution

Amblystogenium minimum is endemic to the eastern Crozet group of islands (Île de la Possession and Île de l'Est) (Vernon and Voisin 1990; Dreux and Voisin 1992; Voisin *et al.* 2017). It was only discovered in 1968 by Lewis Davies (Luff 1972).

Identification

A. minimum is 5–7 mm long and black in color. The elytra have three large erect bristles (clearly visible in the lateral view). The abdomen slightly exceeds the elytra (Figure 5.36). Its silhouette is very flat in the lateral view.

Figure 5.36. Amblystogenium minimum *(photo: B. Chaubet)*

Ecology

A. minimum lives in the fellfield, possibly with *A. pacificum*. It rarely descends into the lower parts covered with vegetation, unlike *A. pacificum*. It is found under rocks and in small crevices. It feeds on earthworms, insect larvae and woodlice (Davies 1972, 1982; Vernon *et al.* 1999).

5.1.6.2. Amblystogenium pacificum *(Putzeys, 1869)*

(Arthropoda; Insecta; Coleoptera; Trechidae)

Distribution

Amblystogenium pacificum is endemic to the Îles Crozet. It was discovered by a whaler and described as early as 1869 (Dreux 1965b). It is present on all the islands of both the eastern and western groups of the archipelago (Jeannel 1940; Dreux 1966; Vernon and Voisin 1990; Dreux and Voisin 1992).

Identification

A. pacificum is larger than *A. minimum*. It is 6–10 mm long. Adults are dimorphic with a black or brown ("rufino" type) colored body and legs. The elytra each have a row of *five to nine bristles (often seven)* clearly visible in the lateral view. The *abdomen clearly extends beyond the elytra* (Figure 5.37). Its silhouette is very flat in the lateral view.

Figure 5.37. Amblystogenium pacificum: *black and brown morphs (photos: B. Chaubet)*

Ecology

A. pacificum is common. It has a wider habitat than *A. minimum* since it is found in the fellfield, as well as at the beach level (Davies 1972, 1982, 1987; Vernon *et al.* 1999). The proportion of black individuals (47% below 50 m a.s.l.) increases with altitude to reach 88% at over 600 m a.s.l. It is likely that the black color allows a better resistance to cold (Davies *et al.* 2007). The species is predatory of earthworms, small lepidoptera larvae and woodlice (Davies 1972).

5.1.6.3. Merizodus soledadinus *(Guérin-Méneville, 1830)*

(Arthropoda; Insecta; Coleoptera; Trechidae)

Distribution

Merizodus soledadinus is the only carabid beetle on the Îles Kerguelen. It is native to Southern America (Tierra del Fuego and the Falkland Islands) and it is likely to have been introduced to the Îles Kerguelen in 1913, with sheep imported from the Falkland Islands. It was discovered in 1939 by René Jeannel (Jeannel 1940), still confined to its place of introduction, Port-Couvreux. Since then, it has spread along the coast and into inland corridors (Figure 5.38). It is currently one of the most invasive invertebrate species on the Îles Kerguelen. It is limited in its extension by the crossing of high altitude areas such as passes (Laparie 2011). Its extension has been well documented (Tréhen and Voisin 1984; Dreux *et al.* 1992; Laparie *et al.* 2010; Lebouvier *et al.* 2011). It has also been introduced to South Georgia where it was first observed in 1963 (Darlington Jr. 1970).

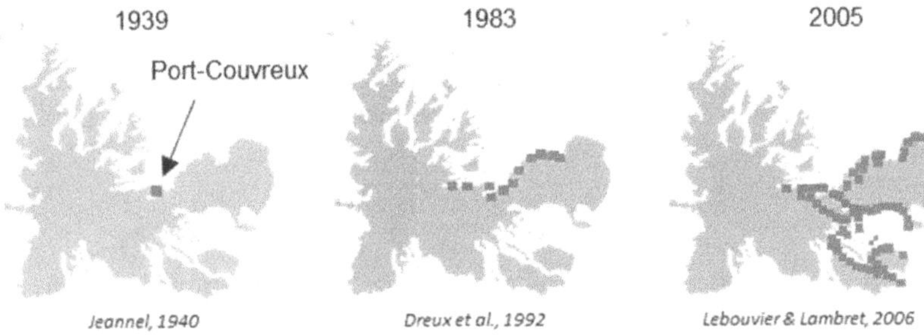

Figure 5.38. *The invasion of the Îles Kerguelen by* Merizodus soledadinus

Identification

M. soledadinus is 5–6.5 mm long. It is dark brown in color, smooth and shiny with lighter legs. Forelegs are very short (Figure 5.39).

Figure 5.39. Merizodus soledadinus *(photo: B. Chaubet)*

Ecology

M. soledadinus lives mainly in coastal areas covered with low vegetation. It can be very abundant beneath pieces of wood, or stones on the foreshore and also where large amounts of decaying organic matter are present. These habitats are home to many prey, such as maggots of native or introduced flies and weevil larvae. Its impact on native fauna may be significant and locally, may even lead to the elimination of certain species such as the wingless fly *Anatalanta aptera* (Chevrier *et al.* 1997; Lebouvier *et al.* 2011; Renault *et al.* 2015; Ouisse *et al.* 2017).

5.1.6.4. Temnostega antarctica *(Enderlein, 1905)*

(Arthropoda; Insecta; Coleoptera; Trechidae)

Distribution

Temnostega antarctica is endemic to the Îles Crozet (Dreux 1971b; Voisin *et al.* 2017). It is one of the numerous species discovered during the *Gauss* expedition in 1902 (Jeannel 1940).

Identification

T. antarctica is a small beetle (3–4 mm long). It is light brown in color, bright and smooth. Its size and color makes it easy to recognize (Figure 5.40).

Figure 5.40. Temnostega antarctica *(photo: B. Chaubet)*

Ecology

T. antarctica lives in damp places. It is found in the vegetation of river banks as well as in mosses, tussocks, under stones and under the roots of plants, provided that it is sufficiently humid. It can also be found in abandoned albatross nests. It is a low altitude species (Jeannel 1940; Dreux 1965b; Dreux 1966; Davies 1973; Dreux and Voisin 1992; Voisin *et al.* 2017).

5.2. Order: Diptera

With 27 species, Diptera is the second largest insect order represented on both the Îles Kerguelen and Île de la Possession (Table 5.3). It is also the most diverse order with 15 families. The proportion of introduced species is higher than for beetles. 40% of the species (11 out of 27 species present in total) are introduced species, compared to 8% for beetles. Among the 15 native species, eight are strictly endemic to the Îles Crozet, six are endemic to the South Indian Ocean Province and one has a sub-Antarctic distribution. Eight of the 15 native species are flightless. They have transformed, reduced or completely absent wings. Flightlessness is probably the most spectacular adaptation of insects to the sub-Antarctic environment.

Family	Species	Îles Kerguelen	Île de la Possession
Anthomyidae	*Fucellia tergina*	Introduced	Introduced
Calliphoridae	*Calliphora vicina*	Introduced	
Canacidae	*Apetaenus enderleini*	South Indian Ocean	South Indian Ocean
	Apetaenus litoralis	South Indian Ocean	South Indian Ocean
Carnidae	*Hemeromyia (?)* sp.[3]	Introduced	Introduced
Chironomidae	*Belgica albipes*		Endemic
	Bryophaenocladius sp.	Introduced	Introduced
	Limnophyes minimus	Introduced	Introduced
	Microzetia mirabilis		Endemic
	Parochlus crozetensis		Endemic
	Smittia sp.	Unknown	
	Telmatogeton amphibius	South Indian Ocean	South Indian Ocean
Ephydridae	*Amalopteryx maritima*	Sub-Antarctic	Sub-Antarctic
Helcomyzidae	*Paractora dreuxi*		South Indian Ocean
Keroplatidae	*Macrocera crozetensis*		Endemic
Micropezidae	*Calycopteryx moseleyi*	South Indian Ocean	
Psychodidae	*Psychoda parthenogenetica*	Introduced	Introduced
Scatopsidae	*Scatopse notata*	Introduced	Introduced

3 This identification of the genus has to be confirmed (see section 5.2.4).

Sciaridae	*Lycoriella sativae*	Introduced	Introduced
	Lycoriella ingenua	Introduced	Introduced
Simuliidae	*Crozetia crozetensis*		Endemic
	Crozetia seguyi		Endemic
Sphaeroceridae	*Anatalanta aptera*	South Indian Ocean	South Indian Ocean
	Anatalanta crozetensis		Endemic
	Leptocera curvinervis	Introduced	Introduced
	Siphlopteryx antarctica		Endemic
Trichoceridae	*Trichocera maculipennis*	Introduced	

Table 5.3. *Diptera of the Îles Kerguelen and Île de la Possession*

5.2.1. *Family: Anthomyiidae*

Anthomyiidae is a large and diverse family, of about 2,000 species of Brachycera flies. One species has been introduced on both islands.

5.2.1.1. Fucellia tergina *(Zellerstedt, 1845)*

(Arthropoda; Insecta; Diptera; Anthomyiidae)

Distribution

The first observations of *Fucellia* were made by René Jeannel in 1939, on the Îles Kerguelen and Île de la Possession (Séguy 1940). First reported as *F. maritima*, the species was renamed *F. tergina* (Evenhuis 1989). In addition, molecular sequencing of the CO1 gene, from 10 individuals collected from different sites on the Îles Kerguelen and Île de la Possession, indicated 100% similarity with the sequences of *F. tergina* (GenBank: KP412987, Leuchtmann and Michelsen (2016)). We thus consider that all mentions of *F. maritima* concern *F. tergina*.

Fucellia tergina is a common species on the marine shores of Europe, North America and the Mediterranean. Since its first record, the species has been regularly observed on the Îles Kerguelen, where it has become largely invasive (Frenot *et al.* 2005). The situation is different on Île de la Possession. Indeed, this species was not re-observed by Philippe Dreux or Lewis Davies during their stays (Dreux 1966; Davies 1973). However, it has been sampled on several occasions since 1990, but always irregularly and in small numbers (Figure 5.41). *F. tergina* is also present on Île Amsterdam and Île Saint-Paul in the South Indian Ocean, and on the Tristan da

Cunha Islands group in the South Atlantic Ocean (Evenhuis 1989; Chown and Convey 2016).

Figure 5.41. *Currently known distribution of* Fucellia tergina

Identification

Females of *F. tergina* are 4.5–6.5 mm long and males are 4–5 mm. It is dark gray in color. The wings carry *one or two spines* on the edge of the costal vein. Halteres are yellowish (Figure 5.42).

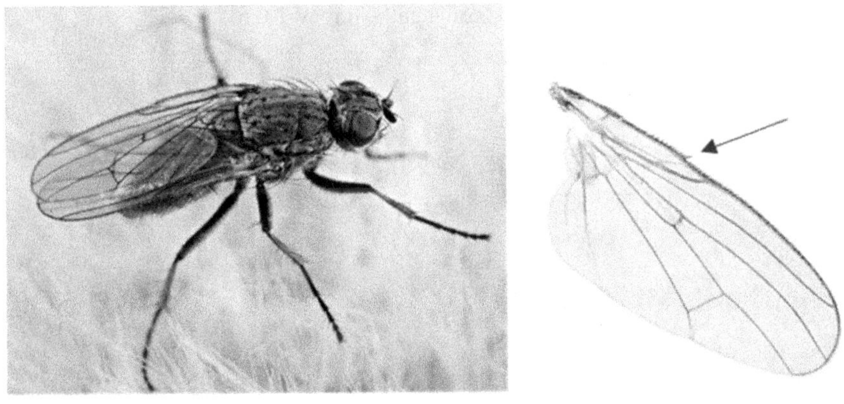

Figure 5.42. Fucellia tergina *female and wing detail (photo: B. Chaubet)*

Ecology

On the Îles Kerguelen, *F. tergina* is very common on the shore. It can abound in decomposing algae. It is a scavenger, feeding on plant debris or dead animals. Its maggots are an important food resource for the introduced ground beetle *Merizodus soledadinus* (Laparie 2011).

5.2.2. *Family: Calliphoridae*

5.2.2.1. Calliphora vicina *(Robineau-Desvoidy, 1830)*

(Arthropoda; Insecta; Diptera; Calliphoridae)

Distribution

Calliphora vicina is a cosmopolitan species found from the tropics to the arctic environments. It appears to have been introduced at least twice to the Îles Kerguelen. The first observation, not followed by establishment, dates from 1902: discovered by Emil Werth in a building at the base (reported under the name *C. erythrocephala* in Enderlein (1908)). It was not re-observed in Port-aux-Français until 1978 by François Xavier de Blignières and Lewis Davies (Chevrier 1996; Chevrier *et al.* 1997). Favored by the global warming of the 1970s, it rapidly colonized new territories, first to the east, and then to the north and south (Frenot *et al.* 2005; Lebouvier *et al.* 2011) (Figure 5.43). The species is also present on Marion Island and in the South Atlantic Ocean Province (the Falkland Islands and South Georgia) (Hänel *et al.* 1998; Chown and Convey 2016).

Figure 5.43. *Currently known distribution of* Calliphora vicina

Identification

C. vicina is 7–12 mm in length. The body is stocky and metallic blue in color. The cheeks are bright orange (Figure 5.44).

Figure 5.44. Calliphora vicina *carrying dandelion pollen (photo: B. Chaubet)*

Ecology

C. vicina is found in almost all biotopes: littoral, herb fields, seabird colonies, fellfield, etc. Adults are active from February to June. They consume nectar and any other sweet materials. This allows them to carry pollen and aid in the seed dispersal of non-native plants, such as dandelion (see Figure 5.44) (Convey *et al.* 2010). After mating, females look for fresh animal cadavers to lay eggs. They compete on baits with native flies like *Anatalanta aptera*. The flying ability of *C. vicina* (unlike flightless native flies), their shorter development time and the larger size of their larvae give advantages in this competition (Chevrier *et al.* 1997; Davies 2006).

5.2.3. *Family: Canacidae*

Canacidae is a small family of about 100 species that live along the coast. Two native species are present on the Îles Kerguelen and Île de la Possession. They have been grouped into the genus *Apetaenus* (Munari 2007). Genus *Apetaenus* is a sub-Antarctic endemic found on Campbell Island, Antipodes Island, Bounty Island,

Macquarie Island, the Îles Kerguelen and the Îles Crozet (Michaux and Leschen 2005).

5.2.3.1. Apetaenus enderleini *(Enderlein, 1909)*

(Arthropoda; Insecta; Diptera; Canacidae)

Distribution

Apetaenus enderleinei is endemic to the South Indian Ocean Province. It is present on the Prince Edward Islands (Marion Island and Prince Edward Island), the Archipel Crozet (Île aux Cochons, Île de la Possession and Île de l'Est) and the Îles Kerguelen (Dreux 1966; Séguy 1971; Crafford *et al.* 1986; Evenhuis 1989).

Identification

A. enderleini is 3.5–4.5 mm long. It is dark in color with *smoke colored wings*. The *first abdominal segments* are indistinct. *Halteres* are yellow (Figure 5.45). Alar polymorphism has been reported by Séguy (1965b).

Figure 5.45. Apetaenus enderleini *(photo: B. Chaubet)*

Ecology

A. enderleini is common on both islands. It is mainly found on the coast and among penguin colonies, as well as in moors and peat bogs. It can be abundant at the seaside in decaying seaweed and kelp deposits (Séguy 1965b; Dreux 1966; Davies 1973; Tréhen and Vernon 1982; Crafford *et al.* 1986). *A. enderleini* is the main host of the parasitoid wasp *Kleidotoma icarus*, which shares the same habitats (Amice 2003).

5.2.3.2. Apetaenus litoralis *(Eaton, 1875)*

(Arthropoda; Insecta; Diptera; Canacidae)

Distribution

Apetaenus litoralis is endemic to the South Indian Ocean Province. The species is present on all the islands except Heard Island and McDonald Island (Dreux 1966, 1978; Gressitt 1970; Crafford *et al.* 1986; Dreux and Voisin 1993). The species was discovered on the Îles Kerguelen during the expedition of the *Challenger* in 1874 and on the Îles Crozet in 1961 by Philippe Dreux (Dreux 1966). A very close species, *A. watsoni* (recently changed to subspecies *A. litoralis watsoni* by Munari (2008)), is present on Macquarie Island and Bishop Island (Greenslade 2006).

Identification

A. litoralis is 3.5–4 mm long. The body is gray and densely hairy. *Wings are reduced* and non-functional. The *first segments of abdomen* are indistinct (Figure 5.46).

Figure 5.46. Apetaenus litoralis *(photo: B. Chaubet)*

Ecology

A. litoralis is a sea-edge species that lives on stranded or decomposing algae, where it can coexist with *Paractora dreuxi* on the Îles Crozet. It is also found in the same coastal Enteromorphs biotopes as those frequented by the midge *Telmatogeton amphibius* (Davies 1973; Dreux and Voisin 1993). Adults are also present on the rocks of the spray zone beside the weevil *Paliroheus eatoni* and the parasitoid wasp *Kleidotoma icarus*, of which it is one of the hosts (Tréhen and Vernon 1982; Mercer *et al.* 2000; Amice 2003). The species feeds on detritus and on macroscopic and microscopic algae such as diatoms (Séguy 1940; Crafford *et al.* 1986).

5.2.4. *Family: Carnidae*

Carnidae is a small family of flies, including 105 species worldwide but mainly found in the holoarctic region (Valera *et al.* 2006; Stuke and Bächli 2015). One species of Carnidae, not yet definitely identified but likely of the genus *Hemeromyia* (R. Meier, pers. comm.), was first observed in 2006 on the Îles Kerguelen and in 2013 on Île de la Possession.

Distribution

Still uncommon on Île de la Possession, the species has colonized several sites on the Îles Kerguelen. It is most often observed in the buildings of Port-aux-Français or in the field huts (Figure 5.47).

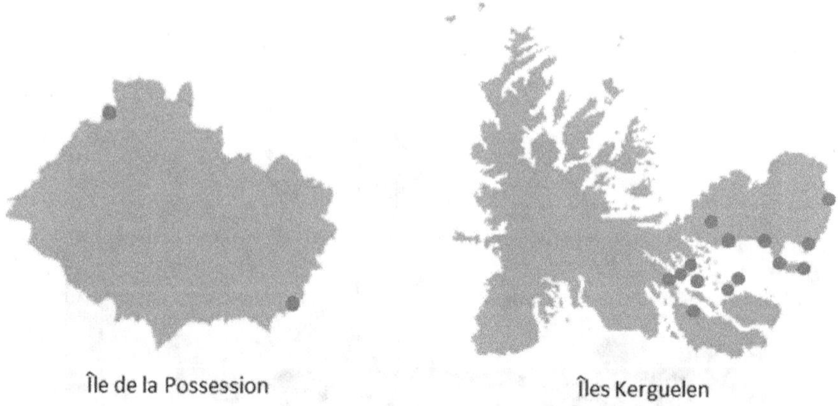

Figure 5.47. *Currently known distribution of* Hemeromyia *sp.*

Identification

This is a small species (2.5–3.5 mm in length), with a metallic gray body, a *head well separated from the thorax* and *whitish halteres*. It differs from a small *Fucellia*, by the *absence of spines* on the costal vein and the venation of the wings (Figure 5.48).

Figure 5.48. Hemeromyia *sp. (photos: B. Chaubet)*

Ecology

Carnids are decomposer flies, coprophagous and necrophagous. Several species are specialized on seabird excrement, which could explain their success on the islands (Brake 1997).

5.2.5. *Family: Chironomidae*

Chironomidae is a large family of Nematoceran flies. Seven species are present on both islands, six on Île de la Possession and four on the Îles Kerguelen. Four species are native and two are introduced. The seventh species, *Smittia* sp., has not yet been identified and its native/alien status is not known.

5.2.5.1. Belgica albipes *(Séguy, 1965)*

(Arthropoda; Insecta; Diptera; Chironomidae)

Distribution

Belgica albipes is endemic to Île de la Possession. It is a rare species, reported only from a few localities and always in very small numbers. It was discovered by

Philippe Dreux in 1962 (Séguy 1965a; Serra-Tosio 1982; Allegrucci *et al.* 2012; Rapp 2013).

Identification

B. albipes is a very small brachypterous chironomid (1.4–1.8 mm long). The body is dark and the legs are long and light yellow (Figure 5.49).

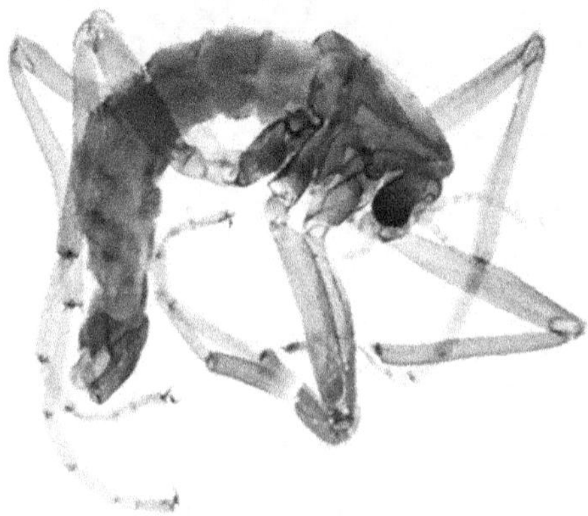

Figure 5.49. Belgica albipes *(photo: B. Chaubet)*

Ecology

B. albipes is an exclusive coastal halophilic species. Larvae live in the puddles on the rocky shore platforms, receiving a supply of fresh water by oozing. They possibly feed on diatoms (Davies 1973; Rapp 2013).

5.2.5.2. Bryophaenocladius *sp.*

Distribution

The genus *Bryophaenocladius* has a cosmopolitan distribution. Species of this genus have been reported from Macquarie Island, the Falkland Islands and Gough Island where it has been introduced (Jones *et al.* 2002; Gaston *et al.* 2003; Greenslade 2006). Its presence on the French islands has only been known since 2017, thanks to an analysis of the CO1 sequences of chironomids, carried out by T. Ekrem and A. Aspaas from the Norwegian University of Science and Technology. We assume that this species has also been introduced there, but its distribution is still incompletely known (Figure 5.50).

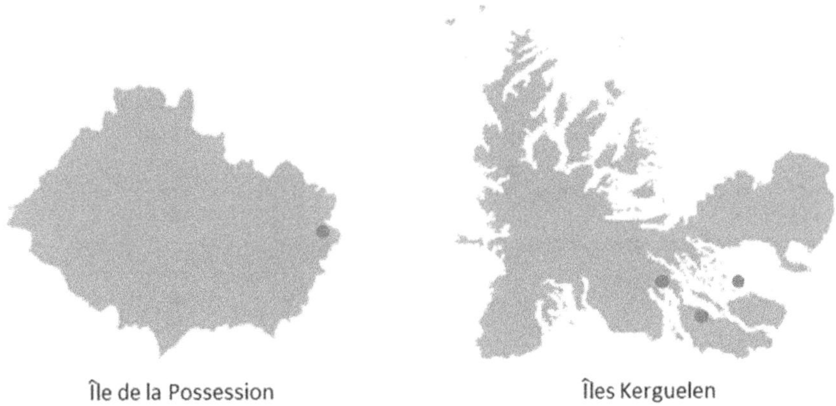

Figure 5.50. *Currently known distribution of* Bryophaenocladius *sp.*

Identification

Bryophaenocladius sp. is a small chironomid less than 1.5 mm long and dark brown in color. It is close to *Smittia* sp. and is distinguished by its *reinforced costal and subcostal veins* (Figure 5.51).

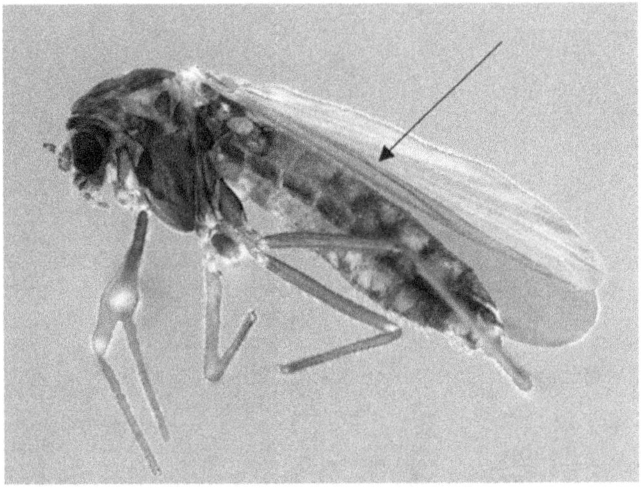

Figure 5.51. Bryophaenocladius *sp. (photo: Aina M. Aspaas, NTNU University Museum CC-BY-NC-SA 4.0)*

Ecology

The ecology of this species is not known. On the Îles Kerguelen, it was sampled along a river bank and in meadows near the coast, and on Île de la Possession within the limits of the Base Alfred-Faure.

5.2.5.3. Limnophyes minimus *(Meigen, 1818)*

(Arthropoda; Insecta; Diptera; Chironomidae)

Distribution

Limnophyes minimus is a holoarctic species which was introduced to many sub-Antarctic islands of the Atlantic Ocean (South Georgia, Gough Island, Nightingale Island, Inaccessible Island) and the South Indian Ocean Province (Prince Edward Island, Marion Island, the Îles Kerguelen, Île aux Cochons, Île de la Possession and Île de l'Est) (Gressitt 1970; Séguy 1971; Saether and Andersen 2011). It was discovered in 1874 on the Îles Kerguelen during the *Challenger* expedition (Tréhen and Delettre 1976) and after 1956 on Gough Island (Gaston *et al.* 2003). Still not very abundant on Île de la Possession, it is considered invasive at Kerguelen, where it has colonized a large number of sites, spread from northwest to southeast (Figure 5.52).

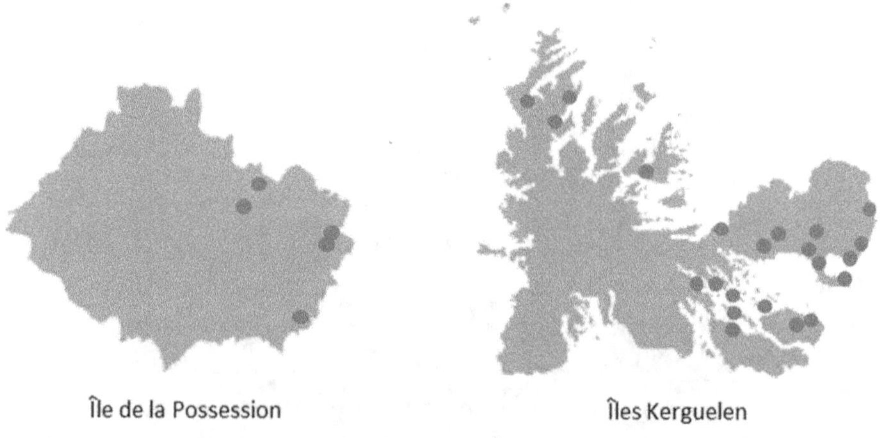

Figure 5.52. *Currently known distribution of* Limnophyes minimus

Identification

L. minimus is a small winged chironomid (1.1–1.9 mm), light brown in color (Figure 5.53).

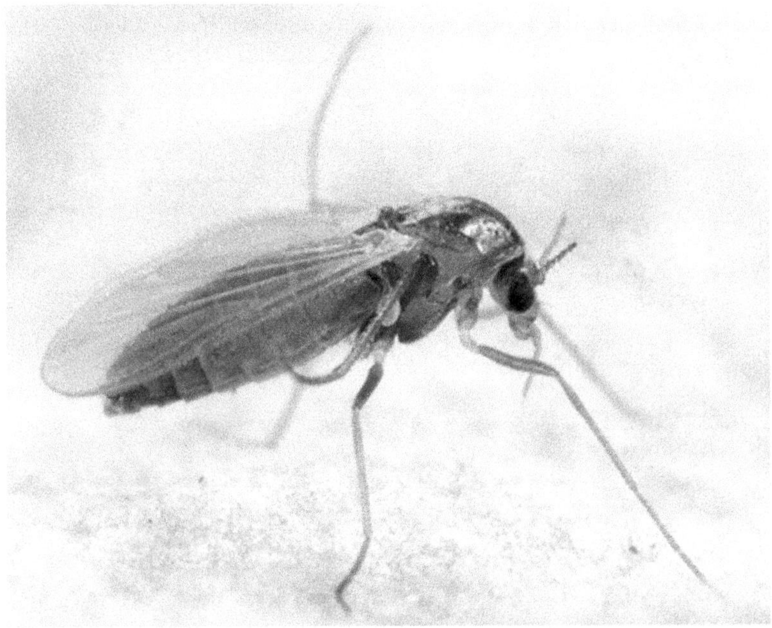

Figure 5.53. Limnophyes minimus *(photo: B. Chaubet)*

Ecology

L. minimus is abundant in wetlands from the banks of floodable rivers to the meadows of *Acaena magellanica*, located on the edges. Larvae feed mainly on plant debris, which is largely decomposed (Delettre 1978; Delettre and Cancela da Fonseca 1979). The species is parthenogenetic (Tréhen and Delettre 1976; Crafford *et al.* 1986). Due to high densities, it probably contributes to litter turnover as much as local species (Hänel and Chown 1998).

5.2.5.4. Microzetia mirabilis *(Séguy, 1965)*

(Arthropoda; Insecta; Diptera; Chironomidae)

Distribution

The species *Microzetia mirabilis* and the genus *Microzetia* are both endemic to Île de la Possession (Séguy 1965a; Serra-Tosio and Brundin 1990).

Identification

M. mirabilis is a small chironomid (1.5–2 mm long). The body is brown and the legs and antennae are orange-yellow. The upper thorax has *large setae* (Figure 5.54). The wings are normally developed but the ratio of the wing length to the body length of 0.7 indicates that the species probably does not fly very well (Tréhen and Vernon 1982).

Figure 5.54. Microzetia mirabilis *(photo: Aina M. Aspaas)*

Ecology

M. mirabilis was first observed on coastal rocks and shore platforms. Since then, it has also been found in torrents further inland (Dreux 1966; Davies 1973). Considered to be rare, we have nevertheless found it several times in large numbers in costal or inland habitats.

5.2.5.5. Parochlus crozetensis *(Serra-Tosio, 1986)*

(Arthropoda; Insecta; Diptera; Chironomidae)

Distribution

Parochlus crozetensis is endemic to Île de la Possession. It was discovered in 1984 by Philippe Vernon in the north of the island (Pointe Basse) (Serra-Tosio 1986).

Identification

P. crozetensis is 1.5–2 mm long. The body is dark brown and the legs light brown. It differs from *M. mirabilis* in the number of segments of labial palps (3 vs. 4) and antennae (10 vs. 11) (Figure 5.55).

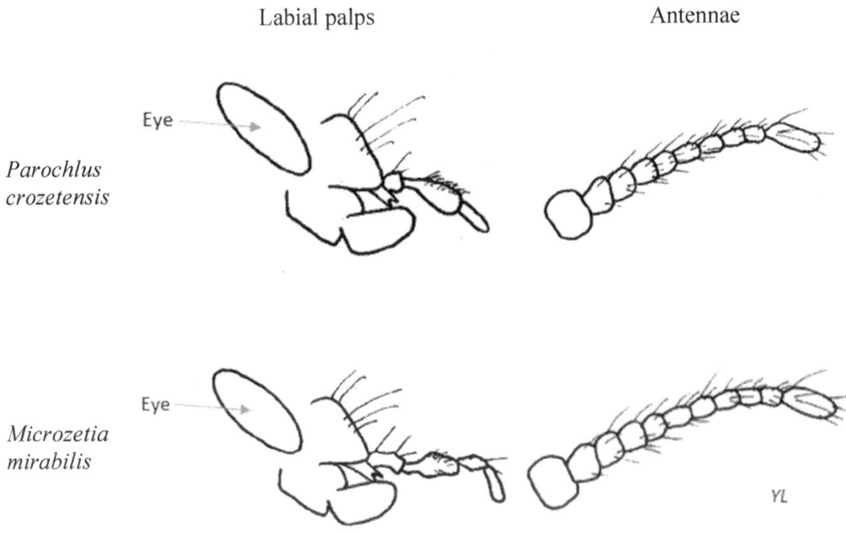

Figure 5.55. *Morphological criteria to distinguish* Parochlus *and* Microzetia *(drawings: Y. Laurent)*

Ecology

P. crozetensis is rare. It was found only in three localities on Île de la Possession, along streams at some distance from the sea (Serra-Tosio 1986; Rapp 2013).

5.2.5.6. Smittia *sp.*

(Arthropoda; Insecta; Diptera; Chironomidae)

Distribution

Smittia is a genus of about 50 species, mostly of European origin. This genus has been reported on several sub-Antarctic islands: Campbell Island, Macquarie Island, Tristan da Cunha Islands and the Îles Kerguelen (Brundin 1962; Michaux and Leschen 2005; Saether and Andersen 2011). On Macquarie Island, one species was probably introduced with sheep (Greenslade 2006). On the other islands, its introduced or native status is not defined. On the Îles Kerguelen, there is supposedly only one species. It is observed in many different sites and environments (Figure 5.56).

Figure 5.56. *Currently known distribution of* Smittia *sp.*

Identification

Analysis of the CO1 sequences indicates that there is *a priori* only one *Smittia* species (BOLD, STIMT152-19...STIMT156-19). It is a small chironomid 1.3 mm long and dark brown in color (Figure 5.57). The antennae have six segments and the labial palps four. It differs from *Limnophyes minimus* (Figure 5.53) mainly in the darker color and the veins, which are difficult to see with the naked eye.

Figure 5.57. Smittia *sp. (photo: B. Chaubet)*

Ecology

Smittia sp. is found in a wide range of environments, such as the coastline (sea leashes, rocky coasts, beaches), riversides, lawns in *Acaena magellanica, Senecio vulgaris* or *Taraxacum officinale*, Kerguelen cabbage (*Pringlea antiscorbutica*), *Poa cookii* tussocks and fellfield areas. Larvae feed on decaying organic material in the soil (Delettre 1983).

5.2.5.7. Telmatogeton amphibius *(Eaton, 1875)*

(Arthropoda; Insecta; Diptera; Chironomidae)

Distribution

Telmatogeton amphibius is endemic to the South Indian Ocean Province. It is present on Marion and Prince Edward Islands, the Îles Kerguelen and the Îles Crozet (Île aux Cochons, Île de la Possession and Île de l'Est). It was discovered during the *Challenger* expedition in 1874 (Séguy 1940, 1971; Dreux 1966; Crafford *et al.* 1986; Chown and Convey 2016). *Telmatogeton* sp. has also been reported from Heard Island (Delettre *et al.* 2003).

Identification

T. amphibius is 3–5.5 mm in length. The *wings are rudimentary* and reduced to an elongated sleeve. The abdomen is slender and gray-brown in color (Figure 5.58).

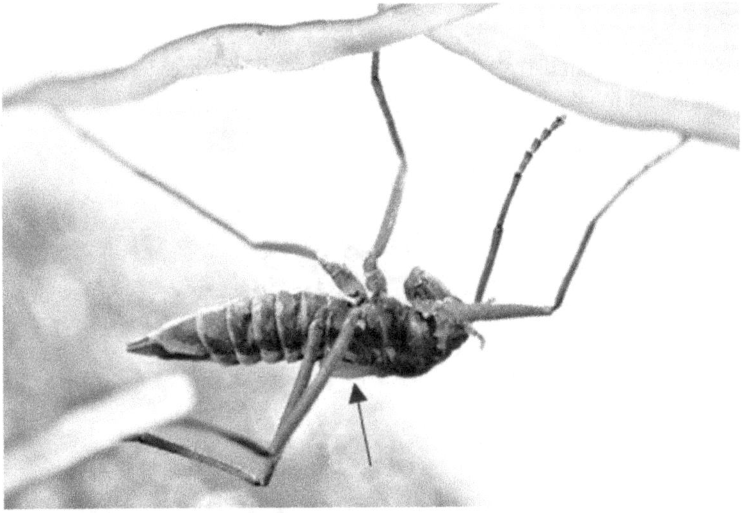

Figure 5.58. Telmatogeton amphibius *(photo: B. Chaubet)*

Ecology

T. amphibius lives on the foreshore in open areas at low tide or in puddles of brackish water (Séguy 1940; Dreux 1966, 1971a; Tréhen and Vernon 1982; Crafford *et al.* 1986). It is quite easily found on *Enteromorpha* green algae. The larvae are euryhaline. Adults are parthenogenetic (Brundin 1962).

5.2.6. *Family: Ephydridae*

Ephydridae is a family of tiny flies that can be found near the seashore or in the vicinity of smaller inland waters, such as ponds. Only one species belonging to the genus *Amalopteryx* occurs on both islands.

5.2.6.1. Amalopteryx maritima *(Eaton, 1875)*

(Arthropoda; Insecta; Diptera; Ephydridae)

Distribution

Amalopteryx maritima is endemic to the South Indian Ocean Province. It is present in the Archipel Crozet (Île aux Cochons, Île des Pingouins, Île de la Possession and Île de l'Est), the Îles Kerguelen and Heard Island (Dreux 1966, 1978; Gressitt 1970). The genus and the species were discovered during the *Challenger* expedition in 1874.

Identification

A. maritima is a small fly (2.5–3.5 mm long), brown in color. The *stenopterous wings* are reduced to sticks (Figure 5.59).

Figure 5.59. Amalopteryx maritima *(photo: B. Chaubet)*

Ecology

A. maritima is a frequent and abundant species on both islands. It mainly frequents the coastal area, sometimes in the company of the flies of the genus *Anatalanta* and *Siphlopteryx antarctica*. Very lively, it moves through the vegetation of *Poa cookii* or *Acaena magellanica* on small sea cliffs, as well as on the bare ground near rookeries or the resting places of elephant seals (Dreux 1971a; De Blignières and Tréhen 1982; Tréhen *et al*. 1985, 1986).

5.2.7. *Family: Helcomyzidae*

Helcomyzidae is a small family of about 80 species. Larvae feed on kelp and other organic matter washed up on the seashore. Only one species is encountered on Île de la Possession.

5.2.7.1. Paractora dreuxi *(Séguy, 1965)*

(Arthropoda; Insecta; Diptera; Helcomyzidae)

Distribution

Paractora dreuxi is endemic to the Prince Edward Islands (Marion Island and Prince Edward Island) and the Archipel Crozet (Îles aux Cochons, Île de la Possession and Île de l'Est) (Gressitt 1970; Séguy 1971; Crafford 1984; Crafford *et al*. 1986). It was discovered in 1964 (Dreux 1965b).

Identification

P. dreuxi is relatively large (6–9 mm long), dark and hairy. The *wings* are very small. The *halteres* are light-colored and clearly visible (Figure 5.60).

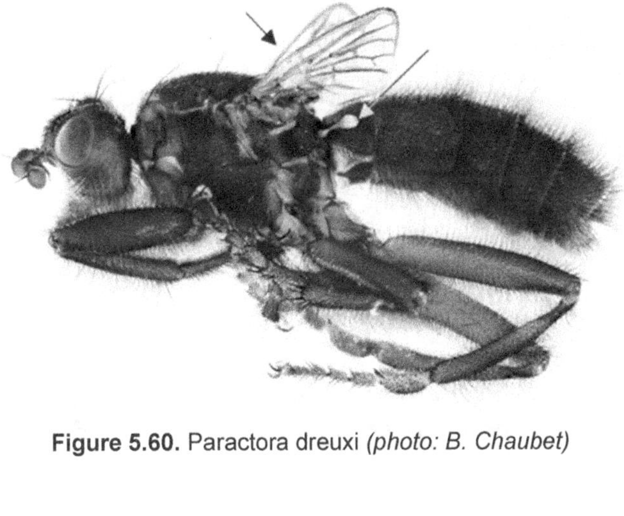

Figure 5.60. Paractora dreuxi *(photo: B. Chaubet)*

Ecology

P. dreuxi is a shore-dwelling species, common among pebbles and may eventually sink deeply into interstitial spaces. It is a primary decomposer of seaweed left on the coast during storms. In summer, when the supply of algae is lower, larvae can consume alternative food, such as marine animal carcasses or even earthworms (Tréhen and Vernon 1982; Tréhen *et al.* 1985; Crafford *et al.* 1986).

5.2.8. *Family: Keroplatidae*

Keroplatidae is a family of at least 950 species mainly living in damp habitats. One species is present in the Archipel Crozet.

5.2.8.1. Macrocera crozetensis *(Colless, 1970)*

(Arthropoda; Insecta; Diptera; Keroplatidae)

Distribution

Macrocera crozetensis is endemic to Île de la Possession. It has been found by Lewis Davies and described in 1970 from only one male (Colless 1970). The female was described five years later (Matile 1975).

Identification

M. crozetensis is 3 mm long. It has an elongated silhouette with legs 5 mm long. The body is brown and the antennae have 15 segments. Wing venation is characteristic with *crossed veins* (Figure 5.61). The females, described by Matile (1975), are brachypterous.

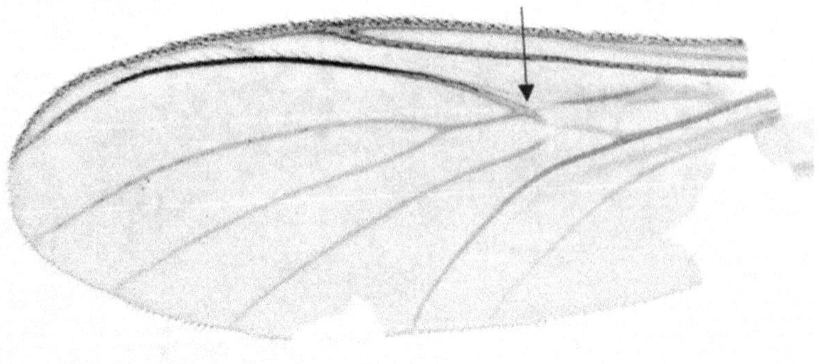

Figure 5.61. Macrocera crozetensis*: male wing (photo: B. Chaubet)*

Ecology

The biology of *M. crozetensis* is poorly known. The species seems rare and we have only observed it once, in the north of Île de la Possession in the vegetation near small ponds. Larvae live buried in earthworm galleries and likely feed on microarthropods (Matile 1975).

5.2.9. *Family: Micropezidae*

Micropezidae is a family of about 500 species in the world, which are particularly well represented in Neotropical regions. Only one species is present on the Îles Kerguelen.

5.2.9.1. Calycopteryx moseleyi *(Eaton, 1875)*

(Arthropoda; Insecta; Diptera; Micropezidae)

Distribution

Calycopteryx moseleyi is endemic to the Îles Kerguelen, Heard Island and McDonald Island (Séguy 1940; Gressitt 1970; Harrison 1970; Evenhuis 1989).

Identification

C. moseleyi is 5.5–6.5 mm long. The wings are rudimentary and reduced to a *small scale*. The halteres are tiny. The body is covered with a grayish-brown pruinosity that gives it a matt appearance. The legs are long with dusky femurs and tibiae (Figure 5.62).

Figure 5.62. Calycopteryx moseleyi *(photo: B. Chaubet)*

Ecology

C. moseleyi is common in the leaf crown of Kerguelen cabbage (*Pringlea antiscorbutica*). Adults move slowly and drop at the slightest alert. Adults lick the surface of the leaves and larvae penetrate the parenchyma on which they feed (Séguy 1940; Dreux 1978; Tréhen *et al.* 1986). Other populations of *C. moseleyi* also grow on stranded algae, particularly in the vicinity of Gentoo Penguin colonies. This habitat is favored because of the reduction of Kerguelen cabbage populations caused by rabbits. These populations have adapted their metabolism to high salt levels (Tréhen *et al.* 1986; Laparie *et al.* 2012; Renault *et al.* 2016).

5.2.10. *Family: Psychodidae*

Psychodidae is a cosmopolitan family of 2,600 small species. Their wings are hairy and held roofwise over the abdomen, giving them a moth-like appearance. This family is represented by only one introduced species on the Îles Kerguelen and Île de la Possession.

5.2.10.1. Psychoda parthenogenetica *(Tonnoir, 1940)*

(Arthropoda; Insecta; Diptera; Psychodidae)

Distribution

Psychoda parthenogenetica has been introduced across the three sub-Antarctic Provinces: the New Zealand Province (Macquarie Island and Campbell Island), the South Atlantic Ocean Province (South Georgia and the Falkland Islands) and the South Indian Ocean Province (Marion Island, Prince Edward Island, the Îles Kerguelen, Île aux Cochons, Île de la Possession and Île de l'Est) (Tillyard 1920; Duckhouse 1970; Gressitt 1970; Vogel and Nicolai 1983; Crafford *et al.* 1986; Evenhuis 1989). *P. parthenogenetica*, with the fly *Fannia canicularis*[4] and the aphid *Rhopalosiphum padi*, is one of the most widespread insects introduced into the sub-Antarctic islands (Frenot *et al.* 2005; Greve *et al.* 2005). On the Îles Kerguelen and Île de la Possession, it has a coastal distribution (Figure 5.63).

---

4 *Fannia canicularis* is not established on the Îles Kerguelen and the Îles Crozet.

Figure 5.63. *Currently known distribution of* Psychoda parthenogenetica

Identification

P. parthenogenetica is 1.7–2.4 mm long. Wings are ovate, pointed and hairy. The veins are arranged in a *radiant pattern*. The segments of the antennae are round and form a structure like a *rosary* (Figure 5.64).

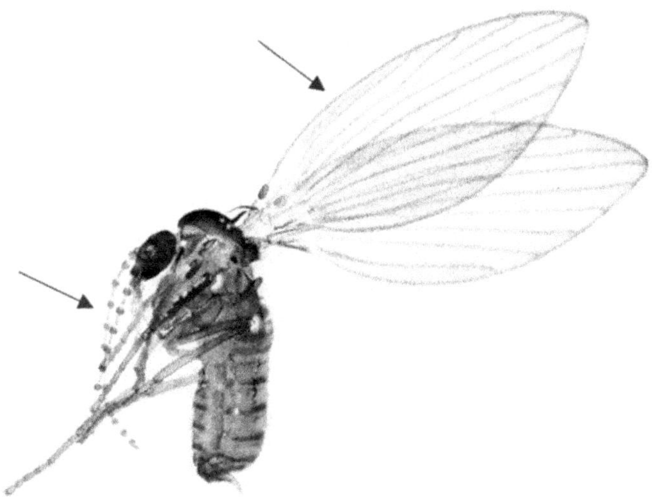

Figure 5.64. Psychoda parthenogenetica *(photo: B. Chaubet)*

Ecology

P. parthenogenetica was almost certainly introduced with vegetables since the establishment of the bases (Davies 1973). The species is found in biotically enriched habitats. The larvae feed on decomposing organic plant or animal matter. They are found in rotting kelp, plant litter such as *Leptinella plumosa*, sewage, animal dung, etc. Adults do not feed because their mouthparts are not functional. They reproduce parthenogenetically (Crafford *et al.* 1986; Hänel 1999; Greenslade 2006).

5.2.11. *Family: Scatopsidae*

Scatopsidae is a small family of around 300 species present all over the world. One species has been introduced to the Îles Kerguelen.

5.2.11.1. Scatopse notata *(Linnaeus, 1758)*

(Arthropoda; Insecta; Diptera; Scatopsidae)

Distribution

Scatopse notata is of palearctic origin and is now worldwide. It has been introduced to the Îles Kerguelen, where it was first observed in 2004, close to a glasshouse at Port-aux-Français and has not dispersed much since (Figure 5.65). The species is also present in South Georgia (Vogel and Nicolai 1983).

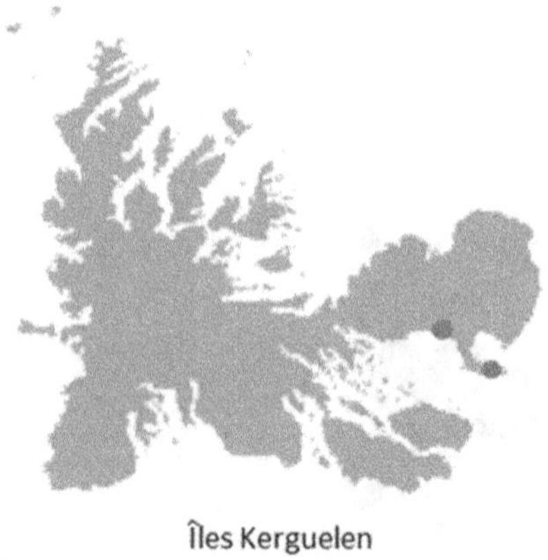

Figure 5.65. *Currently known distribution of* Scatopse notata

Identification

S. notata is 3–3.5 mm long. The wings have a characteristic design with *reinforced costal, subcostal and radial veins*. The abdomen has dark sclerites separated by wide, light bands (Figure 5.66).

Figure 5.66. Scatopse notata *(photo: B. Chaubet)*

Ecology

Initially found in a greenhouse at Port-aux-Français, the species has since been observed in another site far from the base. The larvae develop in decaying organic matter of animal or plant origin (Smith 2013).

5.2.12. *Family: Sciaridae*

Sciaridae is a family of small, dark flies, morphologically very close to each other. Their description in the southern islands is a delicate subject. Several authors have described or recorded three native species on the Îles Kerguelen and one introduced species to the Archipel Crozet (Séguy 1940; Gressitt 1970; Steffan 1972; Crafford *et al.* 1986; Frenot *et al.* 2005), but a recent revision (Broadley *et al.* 2018) has reduced these four species to only two introduced species (Table 5.4). Out of a sample of 30 individuals caught between 2013 and 2017 at 10 remote sites from the Îles Kerguelen and Île de la Possession sent to Adam Broadley (Department of Agriculture, Melbourne, Australia) for identification, only *L. sativae* was identified.

We will therefore limit ourselves here to this single species, even if the presence of *L. ingénue*, reported from Île de la Possession by Steffan (1972) seems likely.

Previous literature			Broadley's 2018 revision
Species	**Status**	**Authors**	
Lycoriella solani	Introduced	Steffan 1972; Frenot 2005	*Lycoriella ingenua*
Sciara womersleyi	Native	Séguy 1940; Gressitt 1970	*Lycoriella ingenua*
Sciara (Bradysia) aubertii	Native	Séguy 1940; Gressitt 1970; Crafford *et al.* 1986; Frenot 2005	*Lycoriella sativae*
Sciara jeanneli	Native	Séguy 1940; Gressitt 1970	*Lycoriella sativae*

Table 5.4. *New synonymy of Sciaridae proposed by Broadley 2018*

5.2.12.1. Lycoriella sativae *(Johannsen, 1912)*

(Arthropoda; Insecta; Diptera; Sciaridae)

Lycoriella sativae is a new synonym for *Sciara aubertii* (Séguy 1940) and *Sciara jeanneli* (Séguy 1940) (Broadley *et al.* 2018).

Distribution

Lycoriella sativae is a holoarctic species which has been introduced to Australia, Hawaï and sub-Antarctic islands (Broadley *et al.* 2018). It is present on the Îles Kerguelen, Île de la Possession and Marion Island (Séguy 1940; Crafford *et al.* 1986) (Figure 5.67). On the Îles Kerguelen, the species has colonized some islands of Golfe du Morbihan and the southeast coast of Péninsule Courbet.

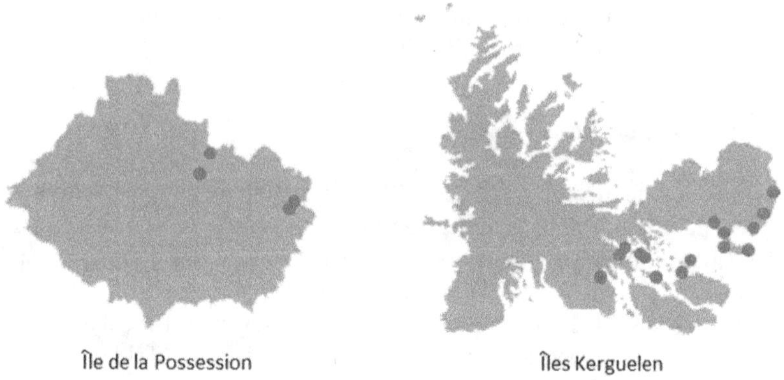

Figure 5.67. *Currently known distribution of* Lycoriella sativae

Identification

L. sativae is 2.3–3 mm long. The body is dark. The family and genus can be identified by the design of the wing, whose M1 and M2 veins form a *characteristic fork* (Figure 5.68). The species is identified and distinguished from *L. ingenua* through the observation of the *male genitalia* (see Figures 5 and 6 in Broadley *et al.* (2018)).

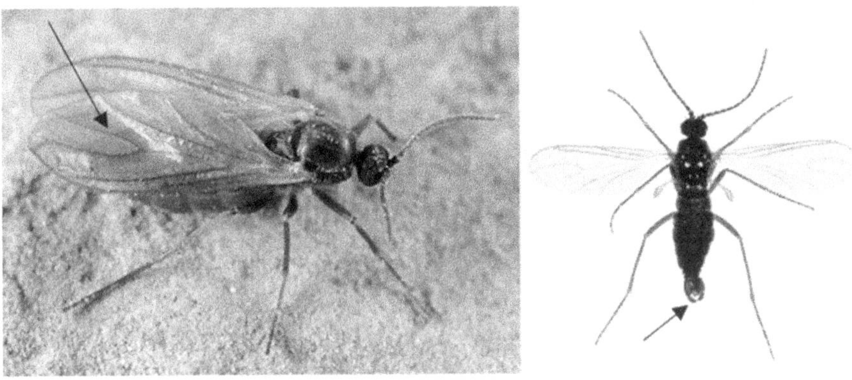

Figure 5.68. Lycoriella sativae, *female (left) and male (right) (photos: B. Chaubet)*

Ecology

The species is common, especially on the Îles Kerguelen, where we have observed it at several sites along the coast, in herbaceous areas as well as on pebble beaches and river banks. Less frequently, we have also found it in seabird burrows. Sciarid larvae feed on fungi and animal feces.

5.2.13. *Family: Simuliidae*

The Simuliidae are found on several isolated volcanic islands that possess suitable running water habitats (Craig *et al.* 2003). This family is represented on the Îles Crozet by two endemic species of the genus *Crozetia*, itself endemic. The species *Crozetia crozetensis*, initially described by Womersley (1937) under the name *Simulium (Cnephia) crozetense*, was found to be a complex of two distinct species, *C. crozetensis* and *C. seguyi*. The male of *C. seguyi* was first described by Beaucournu-Saguez and Vernon (1990). Both species have been recently redescribed by Craig *et al.* (2003).

5.2.13.1. Crozetia crozetensis *(Womersley, 1937)*

(Arthropoda; Insecta; Diptera; Simulidae)

Distribution

Crozetia crozetensis is endemic to Île de la Possession and Île de l'Est (Séguy 1940; Dreux 1966; Craig *et al.* 2003). It was discovered during the BANZARE expedition in 1930 (Dreux 1965b).

Identification

C. crozetensis is 5–5.6 mm long. The body is brownish-black. The species is distinguished from *C. seguyi* by the hairiness of the *katepisternum* (sternite between foreleg and midleg) (Figures 5.69 and 5.70).

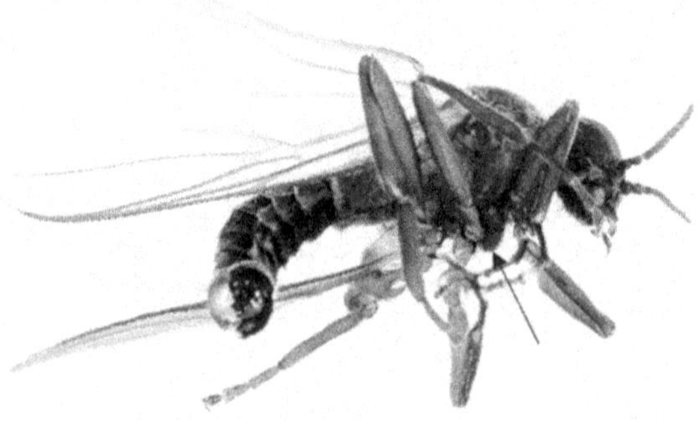

Figure 5.69. Crozetia crozetensis *(photo: B. Chaubet)*

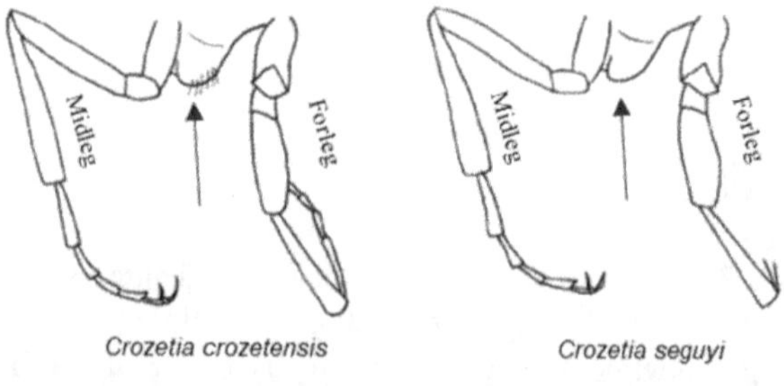

Figure 5.70. *Katepisternum of* Crozetia crozetensis *(hairy) and* Crozetia seguyi *(glabrous) (drawing: Y. Laurent)*

Ecology

C. crozetensis is linked to the presence of streams. Larvae live in torrential waters not loaded in organic matter and browse the river substrates (Craig *et al.* 2003). Adults are non-blood feeding and are mainly observed under stones and in vegetation adjacent to streams (Séguy 1940; Davies 1973, 1974; Tréhen and Vernon 1982).

5.2.13.2. Crozetia seguyi *(Beaucournu-Saguez and Vernon, 1990)*

(Arthropoda; Insecta; Diptera; Simulidae)

Distribution

Crozetia seguyi is endemic to Île de la Possession and Île de l'Est, like *C. crozetensis* (Beaucournu-Saguez and Vernon 1990; Craig *et al.* 2003).

Identification

C. seguyi is 4.5–8 mm long. The body is brownish-black. The *katepisternum* lacks hair (Figures 5.70 and 5.71).

Figure 5.71. Crozetia seguyi *(photo: B. Chaubet)*

Ecology

The ecology of *C. seguyi* is probably similar to that of *C. crozetensis*. It should be noted, however, that, according to our field observations, *C. seguyi* seems to be much more frequent than *C. crozetensis*, which has already been highlighted by Craig *et al.* (2003).

5.2.14. *Family: Sphaeroceridae*

The Sphaeroceridae family includes about 1,300 saprophagous species. Four species are present on Île de la Possession, three native and one introduced. Two of these species are also present on the Îles Kerguelen.

5.2.14.1. Anatalanta aptera *(Eaton, 1875)*

(Arthropoda; Insecta; Diptera; Sphaeroceridae)

Distribution

Anatalanta aptera was discovered by the *Challenger* expedition in 1874. The species is endemic to the South Indian Ocean Province. It is present on the Îles Kerguelen, the Archipel Crozet (Île de la Possession, Île de l'Est, Îlots des Apôtres, Île des Pingouins and Île aux Cochons), Heard and McDonald Islands (Séguy 1940; Dreux 1966; Gressitt 1970; Harrison 1970; Dreux 1978; Vernon 1981; Dreux *et al.* 1988; Vernon and Voisin 1990; Dreux and Voisin 1992).

Identification

A. aptera is a wingless fly, 6.5–7.5 mm in length and dark brown in color. It has a narrow thorax and a particularly *distended abdomen*. This physogastrism is especially remarkable in females (Figure 5.72). It differs from *A. crozetensis* by the *absence of spines* on the femur of the midleg (see Figure 5.73). Some sealers of the 19th Century thought this fly was an ant.

Figure 5.72. Anatalanta aptera, *female (photo: B. Chaubet)*

Ecology

A. aptera is a common species which is sometimes very abundant, especially in seabird colonies. It is found in different habitat types, from sea level up to an altitude of 600 m. It is the only species of fly that remains above 400 m a.s.l. in the fellfield (Vernon 1981). This species is preferably necrophagous, on carcasses of mammals and seabirds. Females can prospect for several hundred meters to find egg-laying sites. The species is also coprophagous on seabird feces and regurgitates, and is even phycophagous on the seabed. It completes its entire life cycle on the soil with seasonal activity, marked by a decrease in food resources during the winter. The adults' abundant lipid reserves allow them to starve for long periods and thus compensate for this seasonality. This particularly long adult life (up to nine months) also compensates for a weak dispersal ability due to flightlessness, by giving them more time to colonize potential egg-laying substrates. On the Îles Kerguelen, it locally suffers from strong predation by the introduced ground beetle *Merizodus soledadinus* and strong competition in the search for egg-laying sites from the introduced fly *Calliphora vicina* (Vernon 1981; De Blignières and Tréhen 1982; Tréhen 1982; Tréhen and Vernon 1982; Tréhen *et al.* 1985, 1986; Chauvin *et al.* 1986; Vernon 1986; Chevrier *et al.* 1997; Lebouvier *et al.* 2011; Greenslade *et al.* 2012).

5.2.14.2. Anatalanta crozetensis *(Enderlein, 1908)*

(Arthropoda; Insecta; Diptera; Sphaeroceridae)

Distribution

Anatalanta crozetensis is endemic to the Archipel Crozet. It is present on Île de la Possession, Île de l'Est and Île des Cochons. It was discovered in 1902 during the stop of the *Deutsche Südpolar-Expedition* at Île de la Possession (Séguy 1940; Dreux 1966; Dreux 1971b; Vernon and Voisin 1990).

Identification

A. crozetensis is very close to *A. aptera*. It is apterous, 6.5–7 mm long and dark brown in color. It is distinguished by the *presence of spines* on the distal end of the femur of the midleg (Figure 5.73).

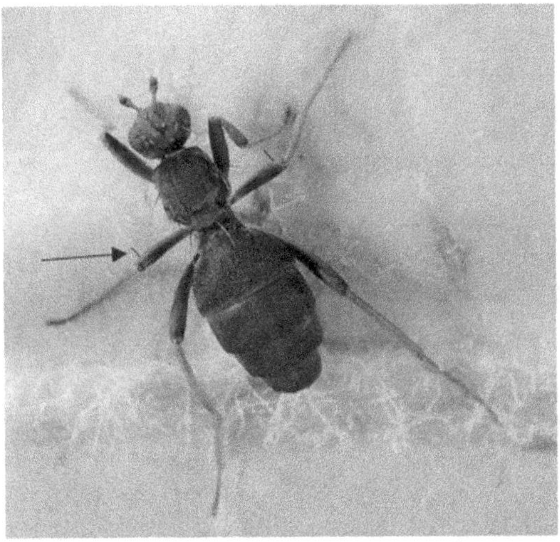

Figure 5.73. Anatalanta crozetensis, *male (photo: B. Chaubet)*

Ecology

The ecology of *A. crozetensis* is similar to that of *A. aptera*. In general, the two species are not found together in the same habitats, with some exceptions, such as rookeries (Vernon 1981). It is also abundant in tussocks and under stones. The coexistence of these two very close species on the same island and, sometimes, in the same biotopes is not yet well explained (Papp 1983; Tréhen *et al.* 1985). The status of *A. crozetensis* as a *bona species* probably deserves to be revised.

5.2.14.3. Leptocera curvinervis *(Stenhammar, 1855)*

(Arthropoda; Insecta; Diptera; Sphaeroceridae)

Distribution

Leptocera curvinervis is a cosmopolitan species found on the Îles Crozet and the Îles Kerguelen. The history of its observations on both Île de la Possession and the Îles Kerguelen is confusing. It was cited as introduced to Île St Paul by Séguy (1940), then as present on the Îles Kerguelen under the name *L. pauliani* by Gressitt (1970) (who himself refers to Séguy 1953), and under the name *L. curvinervis* by Frenot *et al.* (2005). Finally, in their synthesis work, Chown and Convey (2016) cited two species: *L. nigra*, introduced to the Îles Kerguelen and Île St Paul and *Thoracochaeta pauliani*, native to the Îles Kerguelen and Île St Paul. Since 2003, we have rarely observed it on the Îles Kerguelen and only once on Île de la Possession, in 2015 (Figure 5.74). The history of the presence of this species would therefore merit a special re-examination. Pending this revision, we retain the current common name *L. curvinervis*.

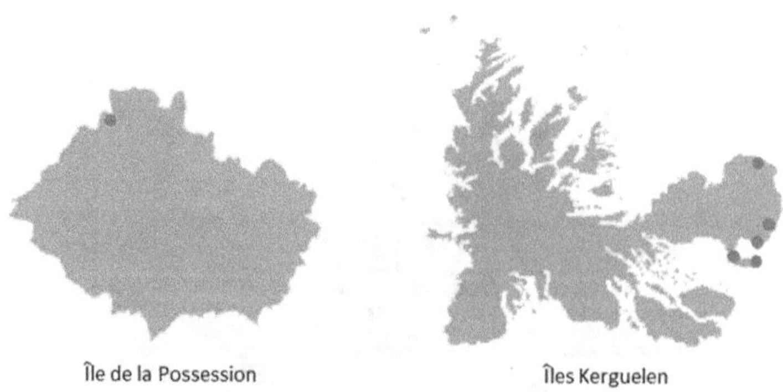

Figure 5.74. *Currently known distribution of* Leptocera curvinervis

Identification

L. curvinervis is 2.3–3 mm long and gray-brown in color. It is easily recognizable by the rounded shape of the wing and its *incomplete venation*, where only the first veins join the opposite edge of the wing. The *costal vein* is reinforced up to half of the wing (Figure 5.75).

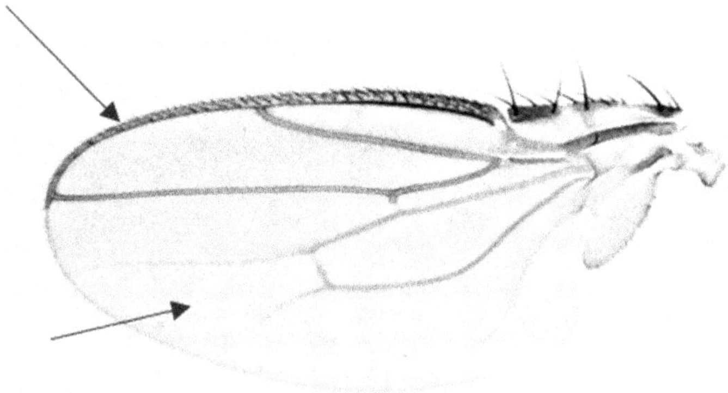

Figure 5.75. *Wing venation of* Leptocera curvinervis *(photo: B. Chaubet)*

Ecology

L. curvinervis is a saprophagous species. Its larvae may develop in seabird burrows where they consume feces and regurgitates.

5.2.14.4. Siphlopteryx antarctica *Enderlein, 1909*

(Arthropoda; Insecta; Diptera; Sphaeroceridae)

Distribution

Siphlopteryx antarctica is endemic to the Archipel Crozet. It is present on all the islands (Séguy 1940; Dreux 1966, 1971b; Vernon and Voisin 1990; Dreux and Voisin 1992). It was also discovered during the 1902 *Deutsche Südpolar-Expedition*.

Identification

S. antarctica is sub-apterous. The wings are reduced to a tiny membranous bulge. This species is 2–3 mm long and grayish-brown in color with a dark abdomen (Figure 5.76).

Figure 5.76. Siphlopteryx antarctica *(photo: B. Chaubet)*

Ecology

S. antarctica frequents the same biotopes as *Anatalanta* sp. It is abundant in environments enriched in organic matter, such as colonies of King Penguins and Gentoo Penguins, as well as in areas with *Poa cookii*. It can exceptionally be found up to 250 m a.s.l. Its small size allows it to explore the environment minutely and probably avoids competition with species with which it shares the same resources, such as *Anatalanta* sp., i.e. seabird feces, feathers and regurgitates (Davies 1973; Tréhen and Vernon 1982; Tréhen *et al.* 1985; Rapp 2013).

5.2.15. *Family: Trichoceridae*

The Trichoceridae family contains about 180 species resembling small crane flies. The genus *Trichocera* includes nine holoarctic species, two of which have been introduced to Maritime Antarctic and sub-Antarctic islands: *T. maculipennis* on the Îles Kerguelen and *T. regelationis* in the South Atlantic Ocean Province (Chown and Convey 2016).

5.2.15.1. Trichocera maculipennis *(Meigen, 1818)*

(Arthropoda; Insecta; Diptera; Trichoceridae)

Distribution

Trichocera maculipennis is a holoarctic species which has been introduced to the Îles Kerguelen. It was first observed by René Jeannel, in 1939, and first described as

Paleopetaurista dubitata by Séguy (1940). It was reclassified as *T. maculipennis* by Dahl (1970). The species was introduced more recently to King George Island in the Maritime Antarctic, where it is now established (Volonterio *et al.* 2013). On the Îles Kerguelen, it is spread around Golfe du Morbihan, near the huts used by scientists and in the vicinity of Port-aux-Français (Figure 5.77).

Figure 5.77. *Currently known distribution of* Trichocera maculipennis

Identification

T. maculipennis has a very thin body and legs. It is 5.5–7 mm long and light grayish-brown in color. The wings have a characteristic venation and pigmentation (Figure 5.78).

Figure 5.78. Trichocera maculipennis *(photo: B. Chaubet)*

Ecology

T. maculipennis is sometimes described as anthropogenic and, indeed, we have almost always observed it in the vicinity of huts. It can also be found in wastewater tanks. Adults can emerge in large numbers at the end of summer days. Larvae are saprophagous and sometimes coprophagous, feeding mainly on decomposing organic matter, such as food waste, decaying plants or algae. Well-adapted to the cold conditions of the sub-Antarctic environment, the species now appears to be well-established (Séguy 1940, Séguy 1953; Smith 2013; Volonterio *et al.* 2013; Hughes and Convey 2014; Huiskes *et al.* 2018).

5.3. Order: Hemiptera

Hemiptera are represented by seven species belonging to two families (Table 5.5). With the exception of *Phthirocoris antarcticus*, an edaphic bug endemic to Île de la Possession, this order includes only introduced species, belonging entirely to the family Aphididae.

Family	Species	Îles Kerguelen	Île de la Possession
Aphididae	*Aulacorthum solani*	Introduced	Introduced
	Macrosiphum euphorbiae	Introduced	Introduced
	Myzus ascalonicus	Introduced	Introduced
	Myzus ornatus	Introduced	
	Myzus persicae	Introduced	
	Rhopalosiphum padi	Introduced	Introduced
Enicocephalidae	*Phthirocoris antarcticus*		Endemic

Table 5.5. *Hemiptera of the Îles Kerguelen and Île de la Possession*

5.3.1. *Family: Aphididae*

Aphididae is a large family, with more than 4,000 species, living mainly in temperate climates. Many aphid species are plant pests. They feed on the sap of plants. This feeding mode did not exist among native insect species, and the introduction of aphids has significantly altered food webs within ecosystems. Six species are present, two of which are invasive. They were very likely introduced by the importation of vegetables and fruit for the use of base personnel as well as on the plants that spontaneously grew on the wooden crates that were previously used to transport the material. Winter climatic conditions allow all species to multiply throughout the year, solely by parthenogenesis. Aphids are polymorphic, having both wingless and winged individuals.

5.3.1.1. Aulacorthum solani *(Kaltenbach, 1843)*

(Arthropoda; Insecta; Hemiptera; Aphididae)

Distribution

Aulacorthum solani is probably of European origin but is now almost worldwide. It has been introduced to Auckland Island, Campbell Island, Antipodes Island in the New Zealand Province, Gough Island in the South Atlantic Ocean Province, Île Amsterdam, Île Saint-Paul, Île de la Possession and the Îles Kerguelen in the South Indian Ocean Province (Gressitt 1964; Palmer 1974; Marris 2000; Gaston *et al.* 2003; Hullé *et al.* 2003a, 2010). On Île de la Possession, the species is present in the Base Alfred-Faure and in some external sites, especially along the east coast. On the Îles Kerguelen, it was first observed in a greenhouse at Port-aux-Français in 1997, and then apparently disappeared following the decommissioning of the greenhouse. It was again observed, from 2005 onwards, on *Acaena magellanica* and *Taraxacum officinale* in the immediate vicinity of the abandoned greenhouse. It has so far remained confined to the base environment (Figure 5.79).

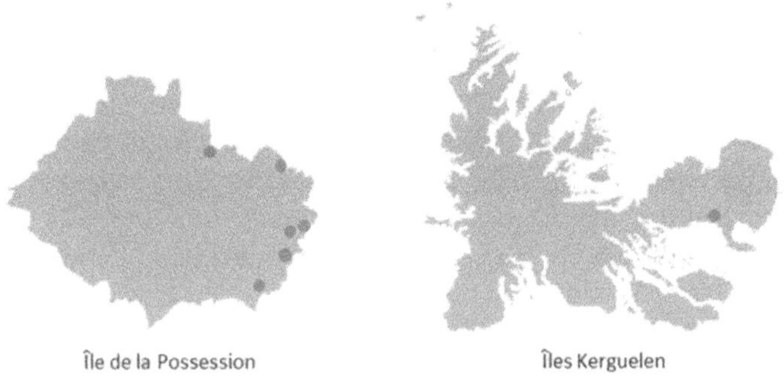

Figure 5.79. *Currently known distribution of* Aulacorthum solani

Identification

A. solani is a fairly large aphid, 2–3 mm long. It is bright green in color. The *cornicles are long, straight and pigmented at the tips*. The alatae are green with darker *transverse streaks* (Figure 5.80).

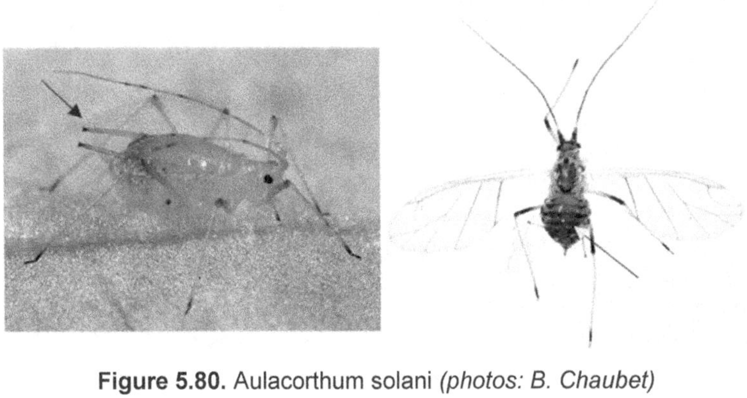

Figure 5.80. Aulacorthum solani *(photos: B. Chaubet)*

Ecology

A. solani is highly polyphagous. It is found on introduced plants (mainly *Taraxacum officinale* and *Cerastium fontanum*) as well as on native ones (*Acaena magellanica, Leptinella plumosa, Pringlea antiscorbutica*) (Hullé *et al.* 2003a, 2003b).

5.3.1.2. Macrosiphum euphorbiae *(Thomas, 1878)*

(Arthropoda; Insecta; Hemiptera; Aphididae)

Distribution

Macrosiphum euphorbiae is probably of North American origin, and is now almost worldwide. It has been introduced on several sub-Antarctic islands: Auckland Island, Campbell Island, the Falkland Islands, Île de la Possession, the Îles Kerguelen and Marion Island (Palmer 1974; Remaudière and Etienne 1988; Hullé *et al.* 2003a; Chown and Convey 2016). The species is also present on Île Amsterdam (Hullé *et al.* 2010). On Île de la Possession, *M. euphorbiae* has been present for several decades but remains uncommon. On the Îles Kerguelen, it was first observed in 2014, on the base, and then, very quickly, in various remote sites (Figure 5.81).

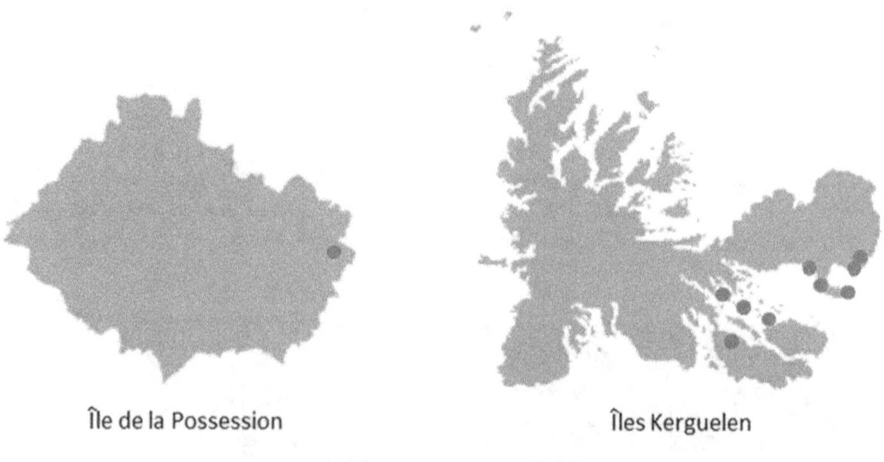

Figure 5.81. *Currently known distribution of* Macrosiphum euphobiae

Identification

M. euphorbiae is a large aphid, 3 mm long. There are green and pink versions. The *cornicles* and *cauda* are very long and straight (Figure 5.82).

Figure 5.82. Macrosiphum euphorbiae *(photo: B. Chaubet)*

Ecology

M. euphorbiae is polyphagous. It feeds on introduced plants (*Senecio vulgaris*, *Taraxacum* sp.) and on native plants (*Acaena magellanica*, *Leptinella plumosa* and *Pringlea antiscorbutica*) (Hullé *et al.* 2003a).

5.3.1.3. Myzus ascalonicus *(Doncaster, 1946)*

(Arthropoda; Insecta; Hemiptera; Aphididae)

Distribution

The origin of *Myzus ascalonicus* is a mystery. It was first observed in 1941 in the United States (Doncaster and Fres 1946), one year later in UK and it is now present from Europe to Australia and New Zealand (Blackman and Eastop 2000). It has been introduced to several sub-Antarctic islands of the New Zealand Province (Auckland Island, Snares Island, Campbell Island, Macquarie Island) and the South Indian Ocean Province (the Îles Kerguelen, Île de la Possession and Marion Island) (Gressitt 1964; Palmer 1974; Crafford *et al.* 1986; Remaudière and Etienne 1988;

Hullé *et al.* 2003a). Its presence on Île de l'Est is likely. The species is also present on Île Amsterdam and Île Saint Paul (Hullé *et al.* 2010). *M. ascalonicus* was first observed on the Îles Kerguelen in the 1960s by Jean-François Voisin. It is nowadays one of the most invasive species on both the Îles Kerguelen and Île de la Possession and the most common aphid species (Figure 5.83).

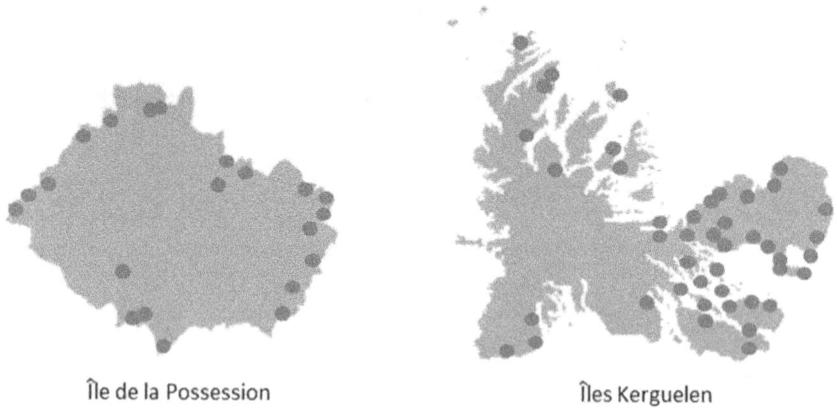

Figure 5.83. *Currently known distribution of* Myzus ascalonicus

Identification

M. ascalonicus is 1.1–2.4 mm long. The apterae are glossy, light brown or yellowish-brown with swollen *cornicles*. The alatae have a *black dorsal patch* (Figure 5.84).

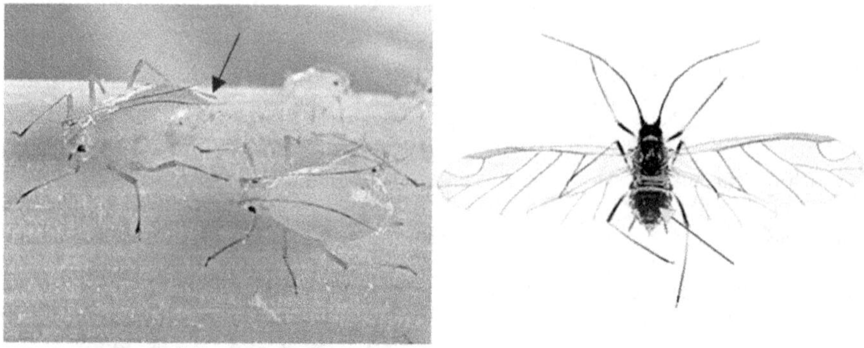

Figure 5.84. Myzus ascalonicus *(photos: B. Chaubet)*

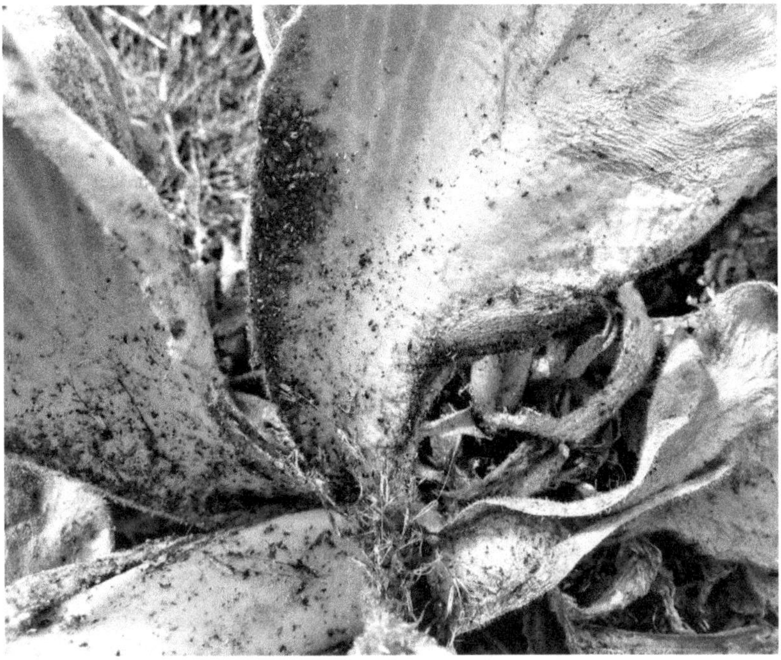

Figure 5.85. Pringlea antiscorbutica*: damage due to* Myzus ascalonicus *(photo: M. Hullé)*

Ecology

M. ascalonicus is highly polyphagous. It multiplies both on introduced plants (*Taraxacum erythrospermum*, *T. officinale*, *Cerastium fontanum*, *Sagina procumbens*, *Senecio vulgaris*, *Stellaria media*, etc.) and on native plants (*Acaena magellanica, Leptinella plumosa* and *Pringlea antiscorbutica*). Its fitness is higher on native plants than on introduced plants, possibly because of the lack of effective defense mechanisms against sap-feeding insects in native plants (Hullé *et al.* 2003b; Hullé 2012). It can cause significant damage, even leading to the decline of the plant (Figure 5.85). It was introduced with its own natural enemies, entomopathogenic fungi belonging to the Entomophthorales order, which severely limit its populations (Papierok *et al.* 2016).

5.3.1.4. Myzus ornatus *(Laing, 1932)*

(Arthropoda; Insecta; Hemiptera; Aphididae)

Distribution

Myzus ornatus is worldwide. It has been introduced to the Îles Kerguelen where it was first observed in 2000. It is also present on Gough Island and Île Amsterdam (Gaston *et al.* 2003; Hullé *et al.* 2003b, 2010) (Figure 5.86).

Figure 5.86. *Currently known distribution of* Myzus ornatus

Identification

M. ornatus is a small aphid, 1–2 mm long. The apterae are light brownish-yellow or matt green with *transverse spots* arranged in pairs on the abdomen. The alatae have a *dorsal dark patch* that is roughly square and sometimes interrupted, on the abdomen (Figure 5.87).

Figure 5.87. Myzus ornatus *(photos: B. Chaubet)*

Ecology

Very polyphagous, *M. ornatus* has so far only been observed on introduced plants, mainly *Taraxacum officinale*, *Senecio vulgaris* and *Cerastium fontanum*. It does not develop large colonies, which probably makes it difficult to detect (Hullé *et al.* 2003a).

5.3.1.5. Myzus persicae *(Sulzer, 1776)*

(Arthropoda; Insecta; Hemiptera; Aphididae)

Distribution

Of Asian origin, *Myzus persicae* is now worldwide. It has been recorded from Auckland Island, Snares Island, Antipodes Island, Inaccessible Island and Gough Island (Palmer 1974; Gaston *et al.* 2002; Chown and Convey 2016). It was introduced to the Îles Kerguelen in 1999 when a greenhouse was recultivated with new plants. It was then regularly observed in the greenhouse until its complete decommissioning in 2009. For the moment, *M. persicae* has not been seen since, but its possible presence in the vicinity of the greenhouse must be monitored.

Identification

M. persicae is 1.2–2.5 mm long. Its color is light green to yellowish-green. The *cornicles* are swollen on their inner side. The alatae have a *dark patch, notched laterally and perforated, on the abdomen* (Figure 5.88).

Figure 5.88. Myzus persicae *(photo: B. Chaubet)*

Ecology

M. persicae can colonize a very large number of plants. On the Îles Kerguelen, it colonized only alien plants in the greenhouse. In temperate regions, it is also the most important virus vector, responsible for severe damage to infected plants. Fortunately, it has remained confined to the greenhouse and has not yet been able to establish outside and cause plant diseases.

5.3.1.6. Rhopalosiphum padi *(Linnaeus, 1758)*

(Arthropoda; Insecta; Hemiptera; Aphididae)

Distribution

Rhopalosiphum padi is of palearctic origin and is now worldwide. It is one of the most commonly-introduced aphid species in the sub-Antarctic islands. It can be found in all provinces: Macquarie Island, Antipodes Island, the Falkland Islands, South Georgia, Marion Island, Prince Edward Island, Île de la Possession and the Îles Kerguelen (Eastop 1962, 1970; Watson 1967; Palmer 1974; Remaudière and Etienne 1988; Hullé *et al.* 2003a). *R. padi* is the second most abundant and common aphid species on both Île de la Possession and the Îles Kerguelen (Figure 5.89).

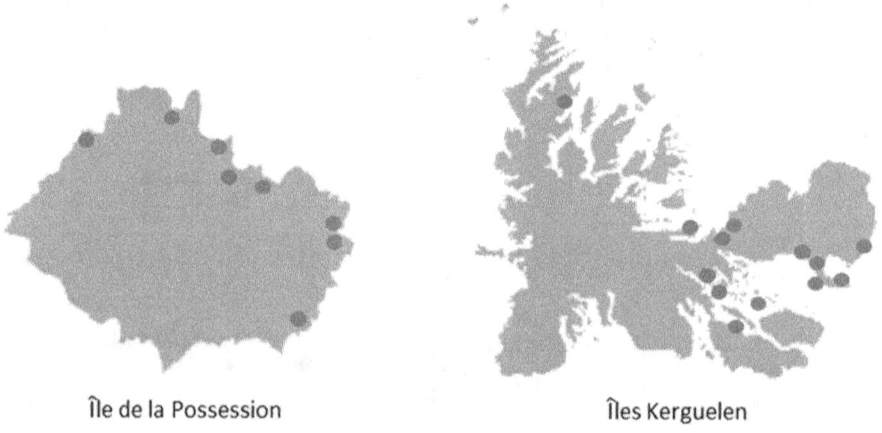

Figure 5.89. *Currently known distribution of* Rhopalosiphum padi

Identification

R. padi is 1.2–2.4 mm long. It is dark green and globular in shape. The alatae have a *reddish spot* around the insertion points of the cornicles. The *cornicles* are short, dark and with a constriction at the tip (Figure 5.90).

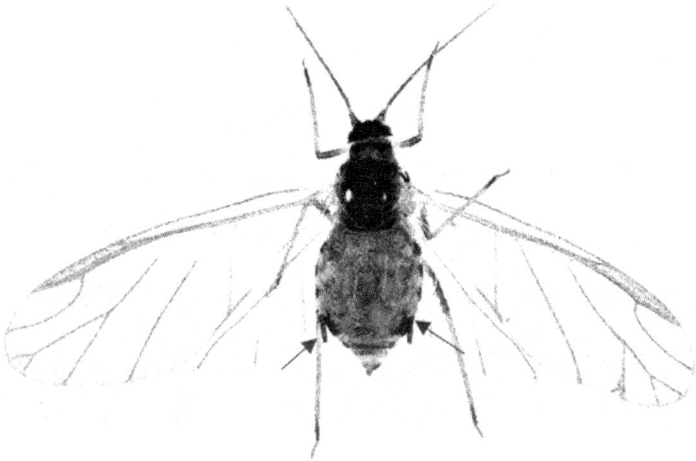

Figure 5.90. Rhopalosiphum padi *(photo: B. Chaubet)*

Ecology

R. padi lives on Poaceae. It has colonized 16 different species, including three native plants. Its main hosts are the native *Poa cookii* and the introduced *Holcus lanatus*, *Trisetum subspicatum* and *Poa annua*. It can be locally very abundant on *P. cookii* and its distribution largely depends on the distribution of this plant (Hullé *et al.* 2003a, 2003b; Lebouvier *et al.* 2011). *R. padi* is also a virus vector. It is most probably responsible for the infection of *P. cookii* with a plant virus, Barley Yellow Dwarf Virus (BYDV). The most likely scenario is that the virus arrived at the same time as its vector aphid and was transmitted secondarily to native plants such as *P. cookii* when the aphid passed from introduced plants to native plants. As *P. cookii* is perennial, it now maintains its own viral population without the need for any transmission by aphids (Svanella-Dumas *et al.* 2013).

5.3.2. *Family: Enicocephalidae*

Enicocephalidae is a small family of about 130 species of bugs. Only one endemic species is present in the Archipel Crozet.

5.3.2.1. Phthirocoris antarcticus *(Enderlein, 1904)*

(Arthropoda; Insecta; Hemiptera; Enicocephalidae)

Distribution

Phthirocoris antarcticus is endemic to the Archipel Crozet. It is present on Île de la Possession, Île de l'Est and Île aux Cochons. It was discovered during the *Gauss* expedition in 1902 (Dreux 1966; Dreux and Voisin 1992).

Identification

P. antarcticus is 3–4 mm long and yellow-brown in color. It is an apterous and neotenic-like bug. It differs from other Hemiptera by being much more slender in body and less sclerotized (Figure 5.91).

Figure 5.91. Phthirocoris antarcticus *(photo: B. Chaubet)*

Ecology

P. antarcticus lives under stones, on the ground, among the roots of tussocks or in Petrel burrows. It is found in heathlands or fellfield from sea level to about 400 m a.s.l. It is a predatory species that probably feeds on collembola, small arachnids and young insect larvae, which it catches with its front legs, which are transformed into tweezers (Dreux 1966).

5.4. Order: Hymenoptera

Hymenoptera is a large order of approximately 230,000 species, including sawflies, bees, wasps and ants. A single native species occurs on both islands.

5.4.1. *Family: Figitidae*

Figitidae is a family of more than 1,400 parasitoid wasp species, including several brachypterous genera.

5.4.1.1. Kleidotoma icarus *(Quinlan, 1964)*

(Arthropoda; Insecta; Hymenoptera; Figitidae)

Distribution

Kleidotoma icarus is endemic to the South Indian Ocean Province. It is present in the Archipel Crozet, the Îles Kerguelen and the Prince Edward Islands. In the Archipel Crozet, it has been reported from the Îles aux Cochons, Île de la Possession and Île de l'Est (Dreux 1966; Quinlan 1967; Crafford *et al.* 1986; Chown and Convey 2016). The species has also been reported from Île des Pingouins but under the name *K. subantarctica*, which is probably a misnomer (Dreux and Voisin 1992). *K. icarus* was officially discovered by Philippe Dreux in 1961 at Île de la Possession (Dreux 1965b), but cynipids had already been reported from the Îles Kerguelen by Patrice Paulian as early as 1953 (Dreux 1966).

Identification

K. icarus is a small wasp (1.5–2 mm long), glossy black, with reduced forewings (Figure 5.92).

Figure 5.92. Kleidotoma icarus *(photo: B. Chaubet)*

Ecology

K. icarus, like the entire subfamily of Eucoilinae to which the genus belongs, is an endoparasitoid of Brachycera flies. It lives on the sea edge in the same habitat as the canacids *Apetaenus litoralis* and *A. enderleini*, its main hosts. It can also parasitize the ephydrid *Amalopteryx maritima* and the small sphaerocerid *Siphlopteryx antarctica*. Larger flies, like *Anatalanta* or *Paractora*, are hardly or not at all parasitized (Davies 1973; Crafford *et al.* 1986; Amice 2003).

5.5. Order: Lepidoptera

The order Lepidoptera includes butterflies and moths. It is a very large order of more than 180,000 described species. Three native species are present on both islands. There are no established alien species (Table 5.6). The order Lepidoptera is also remarkable for the periodic observation of noctuid moths and migrant butterflies. These transient species do not breed on the islands. The most common transient species are presented at the end of the "Lepidoptera" section.

Family	Species	Îles Kerguelen	Île de la Possession
Plutellidae	*Embryonopsis halticella*	South Indian Ocean	South Indian Ocean
Tineidae	*Pringleophaga crozetensis*		Endemic
	Pringleophaga kerguelensis	South Indian Ocean	South Indian Ocean

Table 5.6. *Moths of the Îles Kerguelen and Île de la Possession*

5.5.1. *Family: Plutellidae*

Plutellidae is a small family of phytophagous species, most of which feed on Brassicaceae.

5.5.1.1. Embryonopsis halticella *Eaton, 1875*

(Arthropoda; Insecta; Lepidoptera; Plutellidae)

Distribution

Embryonopsis halticella is endemic to the South Indian Ocean Province. It is present on the Prince Edward Islands (Marion Island and Prince Edward Island), the

Archipel Crozet (Îles aux Cochons, Île des Pingouins, Île de la Possession and Île de l'Est), the Îles Kerguelen and Heard Island (Dreux 1966, 1978; Common 1970; Gressitt 1970). The species was discovered during the *Challenger* expedition, in 1874 (Viette 1948).

Identification

E. halticella is 4–6 mm long and dark brown in color. The *wings* are reduced and cover part of the abdomen, sometimes down to the seventh segment. The abdomen ends with two weakly sclerotized processes called *pseudovalves*. Larvae are pale pink with two darker longitudinal stripes, with the cephalic, prothoracic and the two last abdominal segments sclerotized (Figure 5.93).

Figure 5.93. Embryonopsis halticella *adult and larvae (photos: B. Chaubet)*

Ecology

E. halticella is strictly limited to the tussock grass *Poa cookii*. The adults are active from the early southern summer. They do not feed. The females oviposit on *P. cookii*. The larvae are stem-boring and spend the entire larval stage ensheathed in *P. cookii* tillers. They feed both on the leaves and, in summer, on the inflorescences and seeds. This species is the major herbivore on the Prince Edward Islands, with larvae consuming more than 80 kg/ha of dry mass of *P. cookii* per year (Chauvin and Vernon 1982; Crafford and Scholtz 1986; Crafford *et al.* 1986). The species appears to be more widely distributed on the Îles Kerguelen than on Île de la Possession.

5.5.2. *Family: Tineidae*

Tineidae is a large family of moths, including more than 3,000 species. A very small number of species feed on living plants. The majority, such as the clothes moth and the carpet moth, feed on fungi, lichens and detritus.

5.5.2.1. Pringleophaga crozetensis *(Enderlein, 1905)*

(Arthropoda; Insecta; Lepidoptera; Tineidae)

Distribution

Pringleophaga crozetensis is known only from the Îles Crozet, having been reported from Île aux Cochons, Île de la Possession and Île de l'Est (Chauvin and Vernon 1982; Chown and Convey 2016).

Identification

P. crozetensis is a large species. Adults are 9–12 mm long and late instar larvae can reach 3 cm. Adults are tan-brown in color; the *wings* are very short, reaching just the first abdominal segment. Larvae differ from those of *P. kerguelensis* in that the *dorsal plates* of the thoracic segments are divided into two (Figure 5.94).

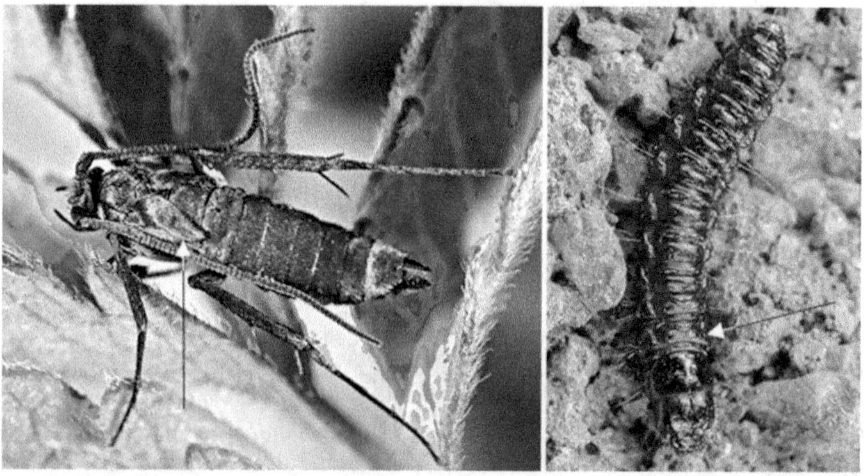

Figure 5.94. Pringleophaga crozetensis *adult and larva (photos: B. Chaubet)*

Ecology

P. crozetensis occurs in different environments, from the shore platforms to the fellfield at an altitude of 900 m a.s.l. with, therefore, a remarkable ability to withstand low temperatures. Its diet has not been described but it can reasonably be assumed to be close to that of *P. kerguelensis*, which is phytophagous and saprophagous (Chauvin and Vernon 1982).

5.5.2.2. Pringleophaga kerguelensis *(Enderlein, 1905)*

(Arthropoda; Insecta; Lepidoptera; Tineidae)

Distribution

Pringleophaga kerguelensis is endemic to the South Indian Ocean Province. It is recorded on Prince Edward Island (but is absent from Marion Island), all the islands of the Archipel Crozet (Île aux Cochons, Îlots des Apôtres, Île des Pingouins, Île de la Possession and Île de l'Est) and the Îles Kerguelen (Dreux 1966; Dreux and Voisin 1992). On Marion Island, it is replaced by *P. marioni* (Viette 1968; Crafford *et al.* 1986). Enderlein's description was based on specimens collected by the *Deutsche Südpolar-Expedition* (Viette 1948).

Identification

P. kerguelensis is the largest flightless insect of the sub-Antarctic islands. It is 14–16 mm long, also tan-brown in color but slightly lighter than *P. crozetensis*. In adults, the *wings* are reduced but reach the middle of the abdomen. Adults have characteristic *orange hairs* on top of their heads. Larvae can reach 50 mm in length. They are distinguished from those of *P. crozetensis* by the *undivided dorsal thoracic plates* (Figure 5.95).

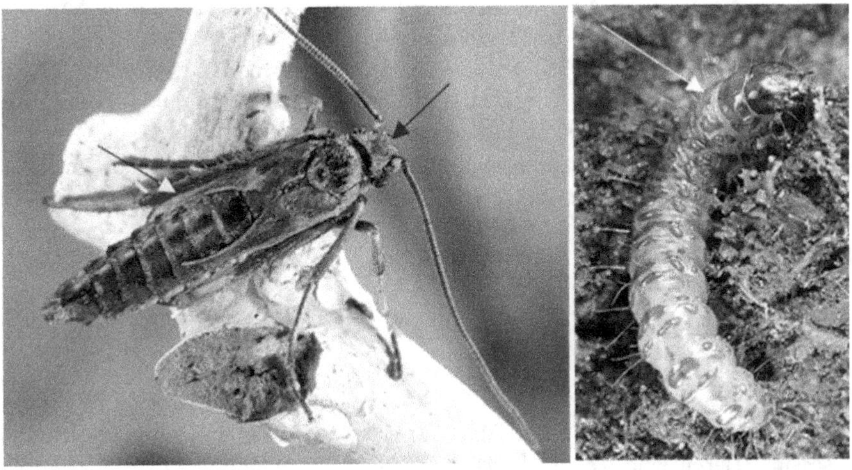

Figure 5.95. Pringleophaga kerguelensis *adult and larva (photos: B. Chaubet)*

Ecology

P. kerguelensis is a common and polyphagous species, which is not only dependent on the Kerguelen cabbage, *Pringlea antiscorbutica*, contrary to what its name might suggest. It is found under the stones of the fellfield, in patches of *Acaena magellanica*, in tussocks, on halophytic lawns and mosses. The larvae are phytophagous and saprophagous. It is a litter-dweller which plays an important role in nutrient turnover (Dreux 1966; Davies 1973; Bellido 1982; Crafford *et al.* 1986; Haupt *et al.* 2014). The larvae are an important food resource for introduced mice (Le Roux *et al.* 2002).

5.5.3. *Transient moths and butterflies*

Several species of Lepidoptera are regularly or occasionally observed on the Îles Kerguelen and Île de la Possession. They are adults and no evidence of reproduction *in situ* has been reported. Some species or individuals probably arrive by boat with fresh or preserved food. Others arrive naturally, carried by the winds during their migrations. We present here the three most frequent species whose observations are due to natural movements.

5.5.3.1. Vanessa cardui *(Linnaeus, 1758)*

(Arthropoda; Insecta; Lepidoptera; Nymphalidae)

Vanessa cardui (Figure 5.96(A)) is a cosmopolitan species, well-known for its long-distance migration. Vagrant adults are regularly reported from the Archipel Crozet, most likely carried by northwest winds from the South African coast, 3,000 km away (Voisin 1975; Rapp 2013). Similarly, *V. cardui* is reported from Marion Island where it is now established (Crafford *et al.* 1986; Chown *et al.* 2002; Greve *et al.* 2017). A very closely related species, *V. kershawi*, is also observed on Macquarie Island during its spring migration from either New Zealand or Australia (Greenslade *et al.* 1999). *V. cardui* does not breed and probably does not feed on Île de la Possession, although sometimes it tries to forage on flowers like *Stellaria media* (Voisin 1975).

5.5.3.2. Agrotis ipsilon *(Hufnagel, 1766)*

(Arthropoda; Insecta; Lepidoptera; Noctuidae)

Agrotis ipsilon (Figure 5.96(B)) is observed very occasionally on Île de la Possession (Barbut and Voisin 2014). *A. ipsilon* is a cosmopolitan species and is a very polyphagous pest. The larvae feed on Asteraceae, Chenopodiaceae, Rosaceae, Brassicaceae and Poaceae. Occasional observations are also reported from other sub-Antarctic islands such as Auckland Island, Campbell Island, Macquarie Island in the New Zealand Province, the Prince Edward Islands in the South Indian Ocean and South Georgia in the South Atlantic Ocean (Crafford *et al.* 1986; Greenslade *et al.* 1999; Marris 2000; Chown *et al.* 2002; Convey 2005). It has not become established on any of the islands except for Marion Island, where the species has been established since 1996 (Chown and Froneman 2008).

5.5.3.3. Plutella xylostella *(Linnaeus, 1758)*

(Arthropoda; Insecta; Lepidoptera; Plutellidae)

Plutella xylostella (Figure 5.96(C)) is also a well-known migratory species with a worldwide distribution. It is reported from several sub-Antarctic islands, such as Marion Island and the Îles Crozet (Chown and Froneman 2008; Rapp 2013), and even arctic islands, such as Svalbard (Coulson *et al.* 2002). It is a major pest of the Brassicaceae. On Marion Island, where it has been established since 1986, it causes damage to the native plant *Pringlea antiscorbutica* (Crafford *et al.* 1986; Crafford and Chown 1987). Its detection in food imported by the supply ships from the French islands led to a ban on the import of fresh cabbage for staff meals.

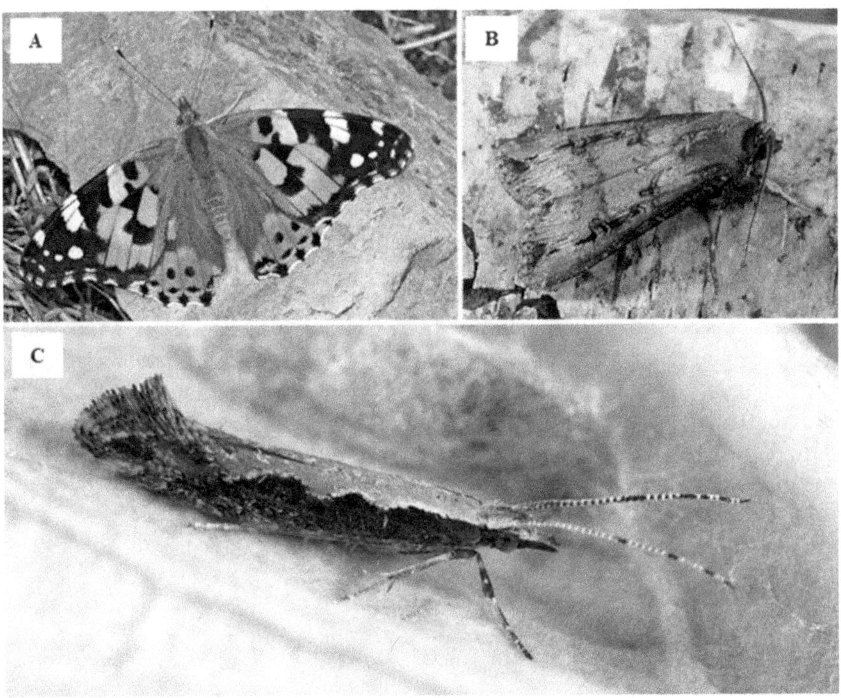

Figure 5.96. *Occasional Lepidoptera species: A)* Vanessa cardui*; B)* Agrotis ipsilon*; C)* Plutella xylostella *(photos: B. Chaubet)*

5.6. Order: Psocoptera

Psocoptera include approximately 3,000 species, feeding primarily on fungi, algae, lichen and organic detritus. Some species also feed on starchy household items, such as grains, paper and even wallpaper glue. From the literature, at least three species should be present on the Îles Kerguelen and Île de la Possession. These are *Antarctopsocus daviesi*, endemic to Île de la Possession, *Rhyopsocus eclipticus*, introduced to the Îles Kerguelen and *Trogium pulsatorium*, introduced to both the Îles Kerguelen and Île de la Possession (Badonnel 1947, 1970). Another two species of the genus *Liposcelis* have also been mentioned: *L. divinatorius* was first observed in 1902 during the *Deutsche Südpolar-Expedition* and was not re-observed by Dreux, who considered it extinct (Enderlein 1908; Dreux 1965b, 1966), and *L. kidder*, cited as native to the Îles Kerguelen by Smithers (1974) and Chown and Convey (2016). During our own investigations over the past 10 years, we found only two morphotypes: one corresponding to *Antarctopsocus daviesi*, but present on both islands, and the other most probably *Lepinotus patruelis*, also present on both islands (Table 5.7).

Family	Species	Îles Kerguelen	Île de la Possession
Elipsocidae	*Antarctopsocus daviesi*	Endemic	Endemic
Trogiidae	*Lepinotus patruelis*	Introduced	Introduced

Table 5.7. *Psocoptera of the Îles Kerguelen and Île de la Possession*

5.6.1. *Family: Elipsocidae*

Within this cosmopolitan family, the genus *Antarctopsocus* was created by André Badonnel (Badonnel 1947) to describe the first species, *A. jeanneli*, found on Marion Island by René Jeannel during the *Bougainville* expedition in 1939, and then on Île aux Cochons in 1964 (Dreux 1965a). Two other genera, *Sandrapsocus* and *Stilopsocus*, also occur on the sub-Antarctic islands of the New Zealand province (Chown and Convey 2016).

5.6.1.1. Antarctopsocus daviesi *(Badonnel, 1970)*

(Arthropoda; Insecta; Psocoptera; Elipsocidae)

Distribution

Antarctopsocus daviesi was considered endemic to the eastern group of the Îles Crozet (Île de la Possession and Île de l'Est) (Badonnel 1970; Chown and Convey

2016). *A. daviesi* was discovered by Lewis Davies in 1968 on Île de la Possession. Very recently, in 2019, it has also been observed on the Îles Kerguelen in Péninsule Courbet. It is replaced by *A. jeanneli* on the western group of the Îles Crozet (Île aux Cochons, Îlots des Apôtres and Île des Pingouins) and on the Prince Edward Islands (Badonnel 1947; Dreux 1966; Crafford *et al.* 1986; Chown and Convey 2016).

Identification

The two species, *A. daviesi* and *A. jeanneli*, are very similar morphologically. They are 2–2.4 mm long and brown in color with dark chocolate-colored eyes. The antennae are 13 segmented, *with segments III, IV and V* being significantly longer than the following ones. The abdomen is lighter, yellow in color. Only extremely reduced *forewings* are present (Badonnel 1970; New 1974; Crafford *et al.* 1986; Lienhard 1998) (Figure 5.97).

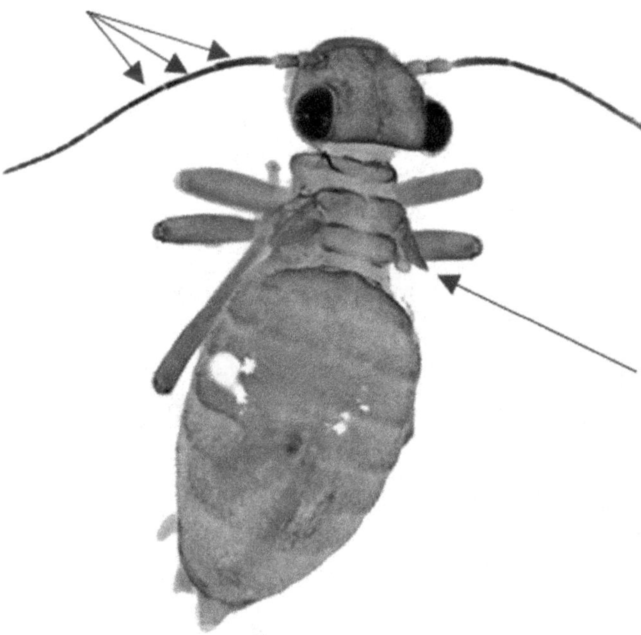

Figure 5.97. Antarctopsocus daviesi *(photo: R. Georges)*

Ecology

Its ecology is probably very close to that of *A. jeanneli*. This species lives from the littoral zone up to 300 m a.s.l. It occurs on rockfaces covered with crustose lichen and also in decayed leaf litter, especially that of the Kerguelen cabbage, *Pringlea antiscorbutica*. It likely feeds on microflora (Crafford and Scholtz 1986).

5.6.2. *Family: Trogiidae*

The species of Trogiidae previously mentioned in the literature from the Îles Kerguelen and Île de la Possession is *Trogium pulsatorium*. We think that it is, in fact, most probably *Lepinotus patruelis* for morphological (see the Identification section below) as well as molecular reasons. The 560 base-pair sequences of CO1 that we obtained for five individuals from the two islands are indeed 99.9% identical to those of *L. patruelis* (GenBank: JQ768777.1, Arif *et al.* (2012)).

5.6.2.1. Lepinotus patruelis *(Pearman, 1931)*

(Arthropoda; Insecta; Psocoptera; Trogiidae)

Distribution

Lepinotus patruelis is present in Europe, North America, Mediterranean countries and New Zealand. It is most probably cosmopolitan (Badonnel 1943; Lienhard 1998; Marris 2000). It has been introduced to the sub-Antarctic islands of Campbell, Snares, Auckland and Antipodes (Smithers 1974; Marris 2000). The species is also present on the Îles Kerguelen and Île de la Possession (Figure 5.98).

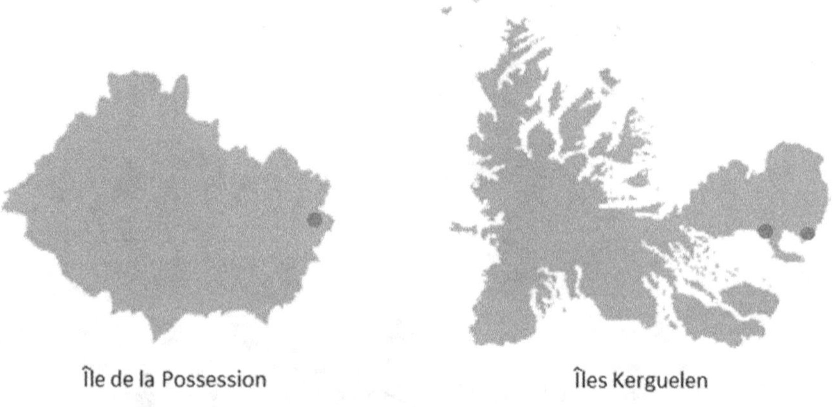

Figure 5.98. *Currently known distribution of* Lepinotus patruelis

Identification

L. patruelis is 2 mm long. It is yellowish-brown with short gray wings. The antennae include more than 20 segments. The species can be distinguished from *T. pulsatorium* by several characteristics: 1) the presence of a *very dark band* between the eye and the base of the antennae; 2) the apical segment of the maxillary palp three times longer than it is wide; and 3) *wings* without venation or reticulum. In addition, the presence of visible *spermatophores* in fertilized females is characteristic of this species (Badonnel 1943; New 1974; Lienhard 1998) (Figure 5.99).

Figure 5.99. Lepinotus patruelis *with one spermatophore (photo: R. Georges)*

Ecology

For the time being, *L. patruelis* seems to be confined to the buildings of the Base Alfred-Faure and Port-aux-Français and a hut on Péninsule Courbet (Kerguelen). It is a free-living scavenger, common in damp, domestic buildings in temperate countries. Its food consists of grain dust, fungi and dead insects (Wearing-Wilde 1996).

5.7. Order: Thysanoptera

Thrips are minute phytophagous insects. There are approximately 6,000 species worldwide. Only one apterous species has been introduced to Île de la Possession and the Îles Kerguelen.

5.7.1. *Family: Thripidae*

Thripidae is the richest Thysanoptera family, with about 2,000 species. They are very specialized, living in the small spaces of the bases of leaves or within flowers.

5.7.1.1. Apterothrips apteris *(Daniel, 1904)*

(Arthropoda; Insecta; Thysanoptera; Thripidae)

Distribution

Apterothrips apteris was previously called *A. secticornis* (Nakahara 1988). This is a cosmopolitan species that has been introduced to several sub-Antarctic islands in the Atlantic and the South Indian Ocean provinces. The species is now present on the Falkland Islands, South Georgia, Marion Island, the Archipel Crozet (Île aux Cochons, Île de la Possession and Île de l'Est), the Îles Kerguelen and Heard Island (Wilson and Stannard 1970; Zur Strassen 1982; Green and Mound 1994; Chown and Convey 2016). Its distribution on Île de la Possession is not well-known and is probably much underestimated. On the Îles Kerguelen, it can be considered invasive, at least in the east (Figure 5.100).

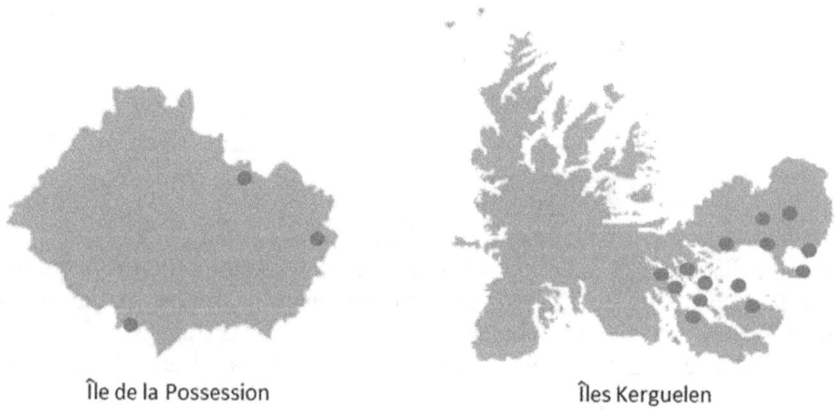

Figure 5.100. *Currently known distribution of* Apterothrips apteris

Identification

A. apteris is the size of a small collembolan (1–1.5 mm). The body is uniformly dark gray with the joints between the thoracic segments and the ends of the legs and antennae lighter (Figure 5.101). Larvae are very similar to adults but yellow.

Figure 5.101. Apterothrips apteris *(photo: B. Chaubet)*

Ecology

A. apteris is a phytophagous species which we have sometimes found in very large numbers in introduced grasslands. It is also present on native plants, such as *Poa cookie*, and on the dicotyledons *Azorella selago*, *Acaena magellanica* and *Crassula moschata*. On Marion Island, it is also invasive and very abundant on *Azorella selago*. It can be found from the coastal lowlands to the fellfield (Davies 1973; Bellido 1982; Zur Strassen 1982; Crafford *et al.* 1986; Barendse and Chown 2001; Greve *et al.* 2017).

6

Originality and Fragility of Invertebrate Fauna

The terrestrial macroinvertebrate fauna of the Îles Kerguelen and Île de la Possession includes a total of 66 native species and 36 established alien species. In this chapter, we first present the main features of the native fauna (biogeography, functional balances and biological adaptations), and then look at the risks posed by introduced species and environmental changes.

6.1. Biogeography and adaptations

6.1.1. *Biogeography*

Remote oceanic islands are perfect places to study evolutionary processes. The observations made by Charles Darwin in the Galapagos Islands or by Alfred Wallace in the Malay Archipelago laid the foundations of geographical aspects of speciation. Among these oceanic islands, the origin of terrestrial species in the sub-Antarctic islands remains an open question. The considerable isolation of the southern Indian Ocean islands has limited the ability of terrestrial invertebrates to colonize them. However, remoteness has not been an insurmountable barrier since many species have passed through it successfully, including poorly mobile species such as earthworms. The Îles Kerguelen and Île de la Possession are currently far from South America, South Africa, Australia and New Zealand but several biogeographical affinities between their fauna and distant potential sources of colonization can be highlighted, even if the colonization scenarios are still often unknown. Here are a few examples of such affinities: (1) the coastal marine genus *Paractora* (Diptera, Helcomyzidae) with the winged *Paractora trichosterna*

(South Georgia, Falkland Islands, Tierra del Fuego) and the flightless *Paractora dreuxi* (Île de la Possession, Marion Island) (Mathis 2011); (2) the large transcontinental harvestman genus *Nuncia* (Opiliones, Triaenonychidae) (61 species) with many New Zealand (and five South American) species and the endemic *Nuncia unifalculata* (Île de la Possession) (Hickman 1939; Porto and Pérez-Gonzalès 2019); (3) the genus *Crozetulus* (Araneae, Anapidae) with the cave-dwelling South African (Cape Province) *Crozetulus scutatus* (Lawrence 1964) and the endemic *Crozetulus minutus* (Île de la Possession) (Brignoli 1981). Similar affinities between mainland and island communities have also been highlighted for plants (Bartish *et al.* 2012; Lehnebach *et al.* 2017).

Such patterns of geographic distribution of species suggest a complex combination of geological and ecological processes. In the footsteps of Charles Darwin, dispersal models, which propose "that present-day distributions were caused by long distance dispersal among disjunct localities" (Endler 1982), were privileged. From the 1960s, the conjunction of cladistics and plate tectonics favored vicariant models which propose "that the distributions resulted from the fragmentation of formerly continuous distributions with accompanying speciation" (Endler 1982). Today, as is often in the history of ecology, a more balanced approach between dispersion and vicariance is most often accepted and a "resurrection of oceanic dispersal in historical biogeography" is even mentioned (de Queiroz 2005). In that context, two hypotheses may still be invoked to explain the origin of these fauna and flora: (1) the vicariance hypothesis based on the post-Gondwana continental drift (e.g. Jeannel (1965)); and (2) scenarios of colonization by dispersal of species after the volcanic emergence of the islands, in the framework of the theory of island biogeography (MacArthur and Wilson 1967). These two approaches differ in the timescales they use, the vicariance hypothesis using older timescales than the post-volcanic emergence colonization scenarios. For the time being, the vicariance hypothesis is more unlikely to hold, as evidence of continental connections is currently controversial. It is possible, however, that several biogeographic histories, involving different hypotheses, may have co-existed in the sub-Antarctic domain to explain the current composition of the fauna. Other factors may also have played a role in explaining the biogeography of these islands; for instance, the Quaternary glaciations and the existence of ice-free refuges, the importance and role of which are still incompletely elucidated.

Moreover, some case studies may lead to difficulties of interpretation. For example, the wingless fly *Anatalanta aptera* (Diptera: Sphaeroceridae) is present on the Îles Kerguelen and Île de la Possession with very weak interindividual morphological variations. It is totally apterous, as described by J.H. Kidder as early as 1876: "The carrion-feeder (*Anatalanta aptera*) has no vestige of either wings or balancers (halteres)" (Kidder 1876). For such flightless flies, when and where did speciation and loss of wings occur? The same difficulties of interpretation arise for other flightless flies present in the two archipelagos, such as *Apetaenus litoralis* and *Amalopteryx maritima,* and for flightless moths such as *Embryonopsis halticella* and *Pringleophaga kerguelensis*.

6.1.2. *Taxonomic and functional disharmony*

Two levels of disharmony must be emphasized in the structure of sub-Antarctic terrestrial macroinvertebrate communities of native species.

Firstly, there is a taxonomic disharmony that is expressed at two levels: (1) of the six insect orders present on both islands (Table 6.1), the two hemimetabolous orders (Hemiptera and Psocoptera) are represented by only two species, and only on Île de la Possession. It may be postulated that the regular production of a large mass of cuticle during molting is metabolically costly in cold environments, a possible obstacle to the establishment of these species in the sub-Antarctic islands (Vernon *et al.* 1998); and (2) on Île de la Possession, among the four holometabolous orders, the predominance of Coleoptera (22 species) and Diptera (14 species) is remarkable in comparison to the very low number of species of Hymenoptera (one species) and Lepidoptera (three species). The same pattern is also observed on the Îles Kerguelen but is less pronounced. It must also be emphasized that nine macroinvertebrate orders are encountered on Île de la Possession but only five on the Îles Kerguelen (absence of Opiliones, Pseudoscorpiones, Hemiptera and Psocoptera). This under-representation of Hemimetabola and the taxonomic dominance of Coleoptera and Diptera is a common feature of all the sub-Antarctic archipelagos of the South Indian Ocean (Table 6.2). Moreover, within a given order, several families are represented on Île de la Possession (Coleoptera: Trechidae, three species; Diptera: Simuliidae, two species; and Helcomyzidae, one species), and absent on the Îles Kerguelen.

Secondly, there is a very pronounced functional disharmony within native arthropod communities, particularly on the Îles Kerguelen. Indeed, in this large archipelago, there are extremely few predatory species (only two spiders), a few phytophagous species (probably partly due to the small number of vascular plants) and a greater number of saprophagous species (probably linked to the importance of resources of marine origin, such as brown algae, feces and molting of seabirds and marine mammals (Vernon *et al.* 1998)). The same pattern is observed on Île de la Possession, except for the greater number of predatory species, with more spiders and three endemic ground beetles.

Île de la Possession

Order	Endemic	South Indian Ocean	Sub-Antarctic	Total
Earthworms	3			3
Mollusks		1		1
Araneids	5	2		7
Opiliones	1			1
Pseudoscorpiones	1			1
Coleoptera	18	3	1	22
Diptera	8	5	1	14
Hemiptera	1			1
Hymenoptera		1		1
Lepidoptera	1	2		3
Psocoptera	1			1
	39	**14**	**2**	**55**

Îles Kerguelen

Order	Endemic	South Indian Ocean	Sub-Antarctic	Total
Earthworms	1			1
Mollusks		1		1
Araneids	1		1	2
Coleoptera	6	5	1	12
Diptera		5	1	6
Hymenoptera		1		1
Lepidoptera		2		2
	8	**14**	**3**	**25**

Table 6.1. *Number of native species according to their level of endemism*

Another important imbalance concerns the number of species in the different archipelagos. The Îles Crozet are home to many more species than the Prince Edward Islands or even the much larger Îles Kerguelen (Table 6.2). This is inconsistent with the Theory of Island Biogeography of MacArthur and Wilson (1967), which predicts that the species richness (balance between immigration of new species and local extinction) of islands depends on their distance from the mainland and on their surface area. The Îles Kerguelen, both older and larger, should therefore host a greater number of species than the Îles Crozet. This apparent paradox of greater diversity (and a particularly high rate of endemism) in the Crozet Islands can be explained by the importance of adaptive radiation processes (this concerns weevils as well as ground beetles, flies and moths) and probably also by a higher survival rate in isolated glacial refuges over long periods of time (Van der Putten *et al.* 2010).

		Prince Edward Islands	Îles Crozet	Îles Kerguelen	Heard and McDonald Islands
Hemi metabola	Psocoptera	1	2	1	-
	Hemiptera	-	1	-	-
Holo metabola	Coleoptera	9	31	12	7
	Diptera	5	14	12	4
	Hymenoptera	1	1	1	-
	Lepidoptera	3	3	2	1
		19	**52**	**28**	**12**

Table 6.2. *Number of native insect species in the four archipelagos of the South Indian Ocean (after Chown and Convey (2016))*

6.1.3. *Biological adaptations*

The remoteness, the cold oceanic climate and the importance of maritime inputs have shaped the natural history of the sub-Antarctic islands. It makes them privileged places also for the study of animal and plant adaptations.

Temperatures in the sub-Antarctic islands are low (annual mean temperature between 4.5 and 5°C) but not extreme, at least on the coast. The lowest temperature ever recorded at sea level was -9.5°C in August 2014, in Port-aux-Français (Météo France data). The oceanic influence limits seasonal variations and maintains low altitude temperatures at levels compatible with the survival of invertebrate species but close to the development thresholds of many of them. In addition, the rapid drop in temperature with altitude (see section 1.1.3) leads to very low temperatures – as low as 200 m a.s.l. (2.5–3°C on average). The result is that the temperature quickly becomes limiting for the development of ectothermic species. Consequently, the majority of macroinvertebrate species live by the sea, in grasslands and wetlands below this altitudinal threshold of 200 m a.s.l. This altitude also corresponds to the upper limit of the area where most of the organic mass of seabirds and marine mammals is concentrated, these marine inputs being an essential trophic resource for a large number of invertebrates. However, not all macroinvertebrates are confined below 200 m a.s.l. Some species such as the earthworm *Microscolex crozetensis*, the weevils *Bothrometopus fasciatus*, *B. gravis* and *Neocanonopsis dreuxi* and the fly *Anatalanta aptera* can live at higher altitudes under very harsh conditions. They then find refuge in protected micro-habitats such as the gaps in the ground, the underside of stones in the fellfield or the cushions of *Azorella selago*.

The small seasonal temperature differences between summer and winter are also an environmental factor to which sub-Antarctic species are adapted. In Port-aux-Français (49°S), the temperature difference between the warmest and the coldest months is 6.1°C for an average annual temperature of 4.8°C. For comparison, at the same latitude under the oceanic climate of the northern hemisphere, seasonal differences are much more marked: for example, in Le Havre (49°N, on the edge of the Channel, in France), the difference is 12.7°C for an average annual temperature of 11.4°C (Météo France data). The challenge for sub-Antarctic ectotherms is therefore to be able to develop in a range of low temperatures present all year round. This combination of low temperature and seasonal stability constitutes an environment which is also found in cave ecosystems (where seasonal variations are even smaller). The biological cycles of native macroinvertebrates are partly the result of these climatic particularities, and several species are active all year round with the possibility, sometimes, of practicing successive laying (iteroparity).

The example of the wingless fly *Anatalanta aptera* (see section 5.2.14.1) is iconic in illustrating these biological adaptations. The adult has an exceptional longevity of around nine months (at 5°C) (Vernon and Vannier 1996). The species is saprophagous and very common in seabird colonies, for example of the Macaroni Penguin (*Eudyptes chrysolophus*). In these often huge colonies (e.g. 100,000 individuals at Cap Cotter, Îles Kerguelen), broken eggs, chick and adult corpses, and feces constitute a very abundant food source for adults and larvae during the summer. But penguins are absent from their colonies for at least five months a year, from May to September. The remarkable fasting performance of the adult flies (up to four months) is very likely linked to this prolonged lack of food resources during the cold months.

The example of *Anatalanta aptera* also helps to illustrate a major adaptation of insects in island environments: the loss of flight function. Flightlessness is a common feature in sub-Antarctic islands, and Jules Verne wrote in *An Antarctic*

Mystery (1897) about the Îles Kerguelen: "A few insects only – butterflies or others – and even these didn't fly, for before they could use their wings, the atmospheric currents carried the tiny bodies away to the surface of the rolling waves". Flightlessness and winglessness affect all the orders of native insects present on both the Îles Kerguelen and the Îles Crozet, where effectively 83% of insects have evolved towards loss of flight or even, for some, outright loss of wings: all Coleoptera, Lepidoptera, Hymenoptera and Psocoptera are flightless. In the Diptera, a gradient of flight ability is observed from winged and flying species (e.g. *Apetaenus enderleini*) to brachypterous and non-flying species (e.g. *Paractora dreuxi*) and finally, to apterous species such as *Anatalanta aptera*. The role of the "wind factor" on islands (as suggested by Charles Darwin) is intuitive but very probably not decisive in sub-Antarctic islands, where the selective advantages of flight ability may have been lost for other reasons. In the case of *A. aptera*, the futility of flight may be due to (a) the energetic costs linked to low temperatures; (b) the low number of predators; and (c) a large amount of trophic resources at certain times of the year, coupled with low interspecific competition. This release of selection pressure for flight resulted, in this species, in the enlargement of the abdomen (physogastrism) and the constitution of lipid reserves replacing flight muscles in the thorax. These two traits are decisive for the high fasting capacity of *A. aptera* during unfavorable periods. It is worth noting that flightlessness and physogastrism are also frequent features in cave flies (Phoridae, Sphaeroceridae) developing in bat guano. When subterranean connectivities are limited, caves may be considered as island-like environments (Mammola 2019). Caves share with sub-Antarctic islands several abiotic and biotic characteristics, the most remarkable of which is perhaps the abundant but temporary presence of bats or seabirds, a possible driver of food resource management by the flies, leading to the accumulation of abdominal reserves and loss of flight.

These examples highlight the value of ecophysiological studies for a better understanding of sub-Antarctic macroinvertebrate ecology.

6.2. Biological invasions and climate change

Native invertebrate species have used natural pathways to colonize the sub-Antarctic islands. These pathways include transport by wind, birds or floating objects (Roderick and Vernon 2009). These colonizing species came from specific areas upstream of the islands. Several authors have thus considered the African continent to have been a probable source of species that naturally colonized the sub-Antarctic islands of the Indian Ocean (see section 6.1.1). This natural process took place over long periods of time, several tens of thousands, or even millions, of years with episodes of regression or, on the contrary, diversification. On the Îles Kerguelen and Île de la Possession, this has resulted in the current fauna being composed of 66 species of native macroinvertebrates. The observation of occasional species, such as *Vanessa cardui*, shows that these natural processes are ongoing.

Since the 18th Century, man-piloted ships have become an additional mode of transport for colonizing species. The potential sources of species have then expanded to the entire globe, depending on the home ports of the ships, and led to the introduction of cosmopolitan species (Frenot *et al.* 2008). Candidate species for immigration must follow a process consisting of four major steps: boarding a supply ship, surviving the journey, coming ashore and, finally, finding a favorable habitat for feeding and breeding. This very selective process leads to an established alien species status, a kind of "green card". To then be qualified as invasive, a new established species must meet two new criteria: 1) disperse beyond its area of introduction and 2) significantly impact its new habitats (Falk-Petersen *et al.* 2006; McGeoch *et al.* 2015). These two processes, dispersion and impact, can take time and depend on climatic conditions in particular. The predatory carabid, *Merizodus soledadinus*, introduced in 1913, remained in the vicinity of its introduction area for several decades before dispersing throughout the southeast of the Îles Kerguelen. Its dispersal accelerated as a result of the global warming that the islands experienced during the 1970s, and considerably increased its impact on prey populations. Likewise, the increase in temperature has allowed the development of the introduced fly, *Calliphora vicina*, which has been greatly favored in its competition with the indigenous flies, by the exploitation of animal corpses. The damage caused to native plants by the aphid *Myzus ascalonicus* depends directly on the size of its colonies. An increase in the mean annual temperature of 1°C (from 5°C to 6°C, for example) would allow the species to complete seven generations per year instead of five currently (Hullé, unpublished data). These three examples show that the role of global warming is therefore all the more important as it increases the number and the impact of invasive species and, at the same time, weakens native species that are highly adapted to a cold and humid environment.

It is difficult to quantify the number of species that go through the first steps of an introduction process before becoming established on the islands. This has been the subject of an assessment for plant propagules by the international program "Aliens in Antarctica" (Huiskes *et al.* 2014). For invertebrates, this is less well known. Thanks to reports from scientific expeditions in the 20th Century and our own more recent observations, it has been possible to compile a list of 22 insect species observed on the islands but never established in a sustainable manner (Table 6.3). Although this list is certainly not complete, it already represents almost as many species as those that have managed to become established. It also includes species belonging to families that currently do not exist on the islands, such as the

Blattidae or the Formicidae. Some species, such as the beetle *Ahasverus advena* and the aphid *Nasonovia ribisnigri*, have been detected recently through the systematic monitoring that has been set up on fresh food products delivered each time the supply ship passes, i.e. four times a year. Those insects introduced but not established are mainly beetles from stored foodstuffs.

On the other hand, the numerous scientific expeditions and ongoing research programs now provide good knowledge on established and invasive species (Table 6.4). 36 species are currently recognized as established on the Îles Kerguelen and Île de la Possession, i.e. 70% of the species introduced in the South Indian Ocean Province (Chown and Convey 2016; Houghton *et al.* 2019). These values would be higher if micro-arthropods, such as mites and springtails, were also taken into account, but knowledge of these groups is not yet sufficient to establish such statistics (Frenot *et al.* 2005). Among the established species of macroinvertebrates, 13 can be qualified as invasive insofar as they have spread widely beyond their site of introduction (see Appendix), they are frequent, and their impact is already visible on the Îles Kerguelen or Île de la Possession, or reported as important on other islands (Soubeyran *et al.* 2015). The omnipresence of the larvae of the blowfly *Calliphora vicina* in the corpses of mammals, and outbreaks of the aphid *Myzus ascalonicus* on native plants such as *Pringlea antiscorbutica* or *Acaena magellanica,* are blatant manifestations of such invasions. This invasive/established species proportion is of the same order as that observed on the Prince Edward Islands (Greve *et al.* 2017).

Diptera and Hemiptera make up the bulk of the invasive insect species. For example, there are very few established Coleoptera species and no Hymenoptera or Lepidoptera, even though these three orders are among the most diverse of insects, especially the Coleoptera. The same taxonomic disharmony is therefore observed in established species as that described for native species. This has not really been investigated, but this disharmony may be partly explained by differences in the ability to survive the journey, according to the development cycles of the species. It is easy to imagine, for example, that a fly larva developing in garbage can more easily survive than an Ephemeroptera larva living in an aquatic environment. Nor can it be excluded that chance played a role in the selection of introduced species.

Introduced species are present in all sectors (except for the central area of Île de la Possession, which seems to be better preserved) (Figures 6.1 and 6.2). The proportion of alien species is higher on the Îles Kerguelen than on Île de la Possession, probably because human frequentation has always been higher on these islands. Similarly, there is a gradient between areas close to the bases, with a higher proportion of alien species and more remote areas, with a lower proportion.

Order	Family	Species	Island	Year	Comments	References
Blattodea	Blattidae	*Blattella germanica* (Linnaeus, 1767)	Kerguelen	1902–1903	Five times in buildings	Enderlein (1908)
Coleoptera	Anthribidae	*Araecerus fasciculatus* (De Geer, 1775)	Possession	2018	In food crates	Anne, pers. comm.
Coleoptera	Coccinelidae	*Lioadalia flavomaculata* (De Geer, 1778)	Possession	1974	Base Alfred-Faure (coming from Cape Town)	Voisin *et al.* (2017)
Coleoptera	Dermestidae	*Dermestes ater* De Geer, 1774	Kerguelen	1930	BANZARE Exp.	Womersley (1937)
Coleoptera	Dermestidae	*Dermestes maculatus* De Geer, 1774	Kerguelen	1902	In buildings, Deutsche Südpolar Exp.	Enderlein (1908)
Coleoptera	Latridiidae	*Corticarina cavicollis* (Mannerheim, 1844)	Possession	2007	In glasshouse	Guthjar and Moncoutier, pers. comm.
Coleoptera	Ptinidae	*Ptinus fur* (Linnaeus, 1758)	Kerguelen	1902	Deutsche Südpolar Exp.	Enderlein (1908)
			Possession	2013	Base Alfred-Faure	Voisin *et al.* (2017)
Coleoptera	Silvanidae	*Ahasverus advena* (Waltl, 1834)	Possession	2007	With fresh food	Guthjar and Moncoutier, pers. comm.
Coleoptera	Tenebrionidae	*Tribolium castaneus* (Herbst, 1797)	Possession	1966, 2008	Stored goods	Voisin *et al.* (2017)
Coleoptera	Trogossotidae	*Tenebriodes mauritanicus* (Linneaus, 1758)	Kerguelen	1902	Deutsche Südpolar Exp.	Enderlein (1908)
Diptera	Calliphoridae	*Calliphora vicina*[1] Robineau-Desvoidy, 1830	Kerguelen	1902	In buildings	Enderlein (1908)
Diptera	Dolichopodidae	*Chrysotus kerguelensis* Enderlein, 1908	Kerguelen	1902	Deutsche Südpolar Exp.	Enderlein (1908)

1 *Calliphora vicina* finally established in 1978, probably after a second introduction (Chevrier 1996).

Order	Family	Species	Island	Year	Comments	References
Diptera	Fanniidae	*Fannia canicularis* (Linnaeus, 1760)	Kerguelen	1874	English Exp. for the Transit of Venus)	Anonymous (1879) and Verrall (1879)
				1898	*Valdivia* Exp.	Enderlein (1903)
Hemiptera	Aphididae	*Brachycaudus helichrysi* (Kaltenbach, 1843)	Possession	2000…	With fresh food	Hullé, pers. obs.
Hemiptera	Aphididae	*Myzus persicae* (Sulzer, 1776)	Kerguelen	1999–2009	Probably disappeared in 2009 after 10 years in a greenhouse.	Hullé *et al.* (2003b)
Hemiptera	Aphididae	*Nasonovia ribisnigri* (Mosley, 1841)	Kerguelen	2000…	With fresh food	Hullé, pers. obs.
Hymenoptera	Formicidae	*Camponotus werthi* Forell, 1908	Kerguelen	1902	In the supplies of the Deutsche Südpolar Expedition from Cape Town (South Africa)	Enderlein (1908) and Forel (1908; 1914)
Lepidoptera	Pyralidae	*Plodia interpunctella* (Hübner, 1813)	Possession	2018	In packages of dry fruit	Anne, pers. comm.
Psocoptera	Liposcelididae	*Liposcelis divinatoria* (Muller, 1776)	Possession	1902	In packaging foam, never seen later	Enderlein (1908) and Dreux (1965b)
Psocoptera	Psoquilidae	*Rhyopsocus eclipticus* Hagen, 1876	Kerguelen	1874	With packing straw by the *Swatara* from Washington (USA)	Hagen (1876) and Eaton (1879)
Psocodea	Trogiidae	*Trogium pulsatorium* (Linneaus, 1758)	Kerguelen Possession	?		Badonnel (1947)
Thysanoptera	Thripidae	*Thrips brachycephala* (Enderlein, 1908)	Kerguelen	1902	In a marine animal colony	Enderlein (1908)

Table 6.3. *List of introduced species that have not become established (names have been updated according to the current taxonomy)*

Phylum/class	Order	Species	Île de la Possession	Îles Kerguelen
Annelida	Lumbriculida	*Allolobophora chlorotica*		Established
		Dendrodrilus rubidus	Established	Invasive
		Eiseniella tetraedra		Established
Mollusca	Stylommatophora	*Deroceras invadens*	Uncertain	Established
		Deroceras reticulatum		Established
Arachnida	Araneae	*Tegenaria domestica*	Synanthropic	Synanthropic
		Lepthyphantes leprosus		Synanthropic
		Ostearius melanopygius	Synanthropic	
		Tenuiphantes tenuis		Established
		Pholcus phalangioides	Synanthropic	
		Steatoda grossa	Synanthropic	Synanthropic
		Steatoda triangulosa	Synanthropic	Synanthropic
Insecta	Coleoptera	*Cartodere nodifer*	Established	Established
		Latridius minutus		Established
		Ptinus tectus		Established
		Merizodus soledadinus		Invasive
	Diptera	*Fucellia tergina*	Established	Invasive
		Calliphora vicina		Invasive
		Hemeromyia (?) sp.	Established	Invasive
		Bryophaenocladius sp.	Established	Established
		Limnophyes minimus	Established	Invasive
		Smittia sp.		Invasive
		Psychoda parthenogenetica	Invasive	Invasive
		Scatopse notata	Established	Established
		Lycoriella sativae	Established	Invasive

Phylum/class	Order	Species	Île de la Possession	Îles Kerguelen
Insecta		*Lycoriella ingenua*[2]	Established	Established
		Leptocera curvinervis	Established	Established
		Trichocera maculipennis		Invasive
	Hemiptera	*Aulacorthum solani*	Invasive	Established
		Macrosiphum euphorbiae[3]	Established	Established
		Myzus ascalonicus	Invasive	Invasive
		Myzus ornatus	Established	Established
		Myzus persicae[4]		Established
		Rhopalosiphum padi	Invasive	Invasive
	Psocoptera	*Rhyopsocus ecliptus*	Established	Established
	Thysanoptera	*Apterothrips apteris*	Established	Invasive

Table 6.4. *List of the introduced macroinvertebrate species*

In contrast to mammals, established and invasive invertebrate species are virtually impossible to destroy (Frenot *et al.* 2008; Convey 2011; Greve *et al.* 2017). Thus, protection measures focus on biosecurity by preventing the introduction of new species and the inter-site dispersal of already established species. Ships, passengers, cargo and fresh products are the only pathways for the introduction of non-native species to the islands. Passengers must thoroughly clean their luggage and personal belongings before going ashore (Figure 6.3) (Huiskes *et al.* 2018). All cargo and fresh products are stored in transit halls and inspected within two days, before being dispatched to the various buildings on the base (Hughes *et al.* 2011). Similarly, clothing and personal equipment are systematically cleaned before moving from one site to another.

Despite all of these controls, some species regularly slip through the control procedures. For example, aphids such as *Nasonovia ribisnigri* and moths such as *Plodia interpunctella* are regularly observed (Table 6.3) but have not established. The early detection of these species is based on the knowledge of specialists, capable of spotting and identifying them. Given the diversity of possibilities, it is not always easy to find experts for each species. Collaboration between scientists and managers

2 The establishment of *Lycoriella ingenua* has to be confirmed.

3 *Macrosiphum euphorbiae* is currently expanding on the Îles Kerguelen.

4 *Myzus persicae* is probably declining or extirpated.

is particularly important to establish correct diagnoses on the identity and status of these species (Hughes and Convey 2012).

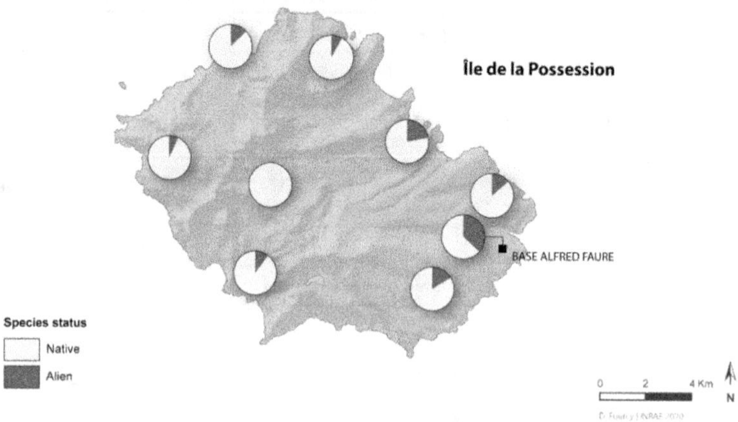

Figure 6.1. *Ratio of alien species on Île de la Possession*

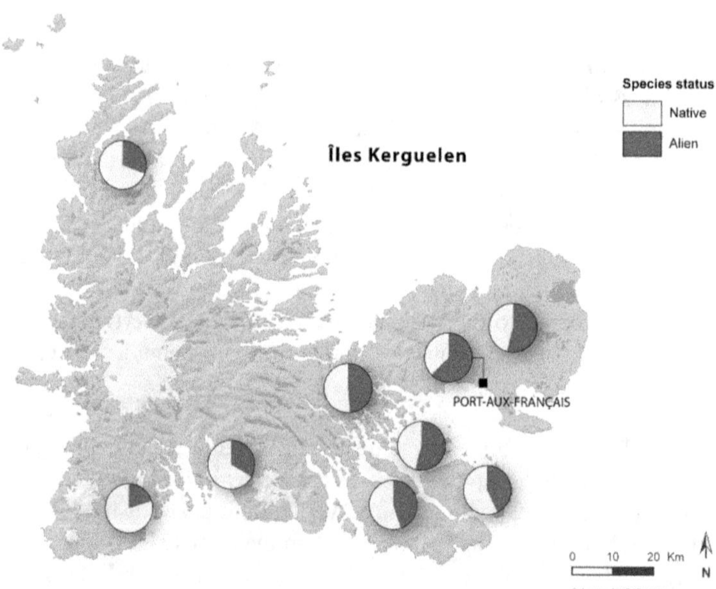

Figure 6.2. *Ratio of alien species on the Îles Kerguelen*

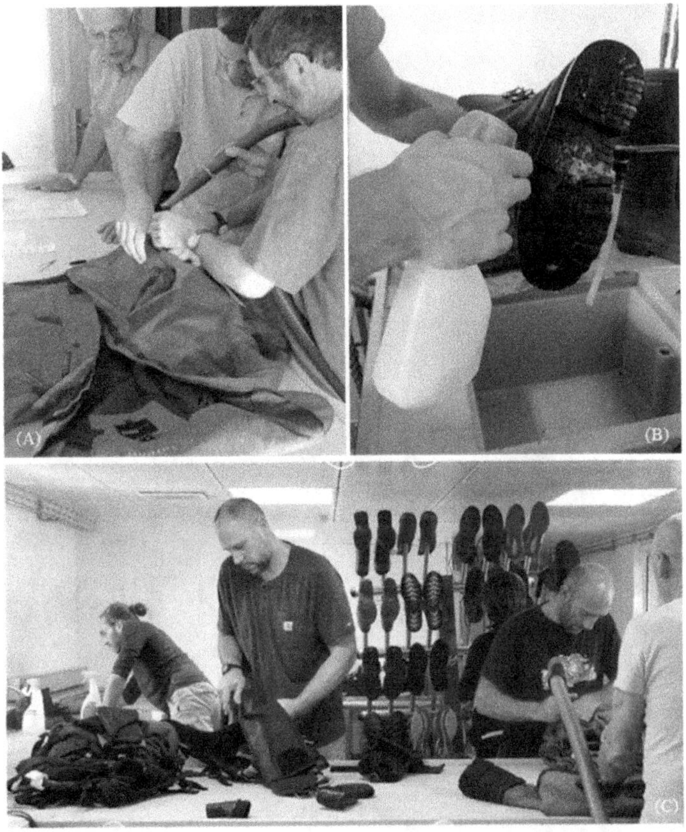

Figure 6.3. *Biosecurity measures before landing (photos: (A) M. Hullé, (B) T.A.A.F. and (C) A. Kiesel)*

Due to increasingly less restrictive climatic conditions and increasing human displacement, it is more than likely that the number of alien species will continue to grow on the sub-Antarctic islands (Chown and Convey 2016; Duffy *et al.* 2017). These prospects are enough to convince us that preventive measures must be maintained and even further improved.

Introductions result in an increase in the number of species and in the complexity of food webs. Beyond important immediate consequences, such as the local disappearance of native species populations, introductions can also allow the progressive integration of species into sub-Antarctic ecosystems and thus fulfill the ecological functions of primitive ecosystems (Keulartz and Van Der Weele 2009). What is a tragedy for the natural heritage can be mitigated by the appearance of new ecological balances.

FIG. 2.—Zoological Laboratory on the Main Deck.

The Zoological Laboratory aboard the H.M.S. Challenger *which visited Prince Edward Island, the Îles Crozet, the Îles Kerguelen and Heard Island in 1874 with the scientists Charles Wyville Thomson, John Murray, Henri Nottidge Moseley, Rudolf von Willemoes-Suhm and the artist John James Wild (after Wyville Thomson and Murray (1885))*

Appendix

Currently Known Distribution of Established Alien Species

This appendix brings together the currently known distribution maps of all the introduced species. Each species is illustrated by a box showing, where appropriate, Île de la Possession on the left and the Îles Kerguelen on the right.

Among these species, some have remained confined to their place of introduction: this is the case for synanthropic species, such as the spiders of the genus *Tegeneria* or *Steatoda*. Other species probably took advantage of human transport to establish in other regularly frequented sites, such as the fly *Leptocera curvinervis* or the thrips *Apterothrips apteris*, found in the vicinity of the huts of Pointe Basse, La Pérouse or Baie Américaine on Île de la Possession. Other species have begun to disperse but are not considered invasive, such as the beetle *Cartodere nodifer*, the earthworm *Eiseniella tetraedra* or the aphid *Macrosiphum euphorbiae*, because their abundance and impact remain low. Finally, 13 species (marked with an asterisk) are currently considered invasive on at least one of the two islands: an earthworm (*Dendrodrilus rubidus*), a beetle (*Merizodus soledadinus*), eight diptera (*Fucellia tergina, Calliphora vicina, Hemeromya* sp., *Limnophyes minimus, Smittia* sp., *Psychoda parthenogenetica, Lycoriella sativae, Trichocera maculipennis*), two aphids (*Myzus ascalonicus, Rhopalosiphum padi*) and one Thrips (*Apterothrips apteris*).

Annelida Crassiclitellata

Lumbricidae

Allolobophora chlorotica

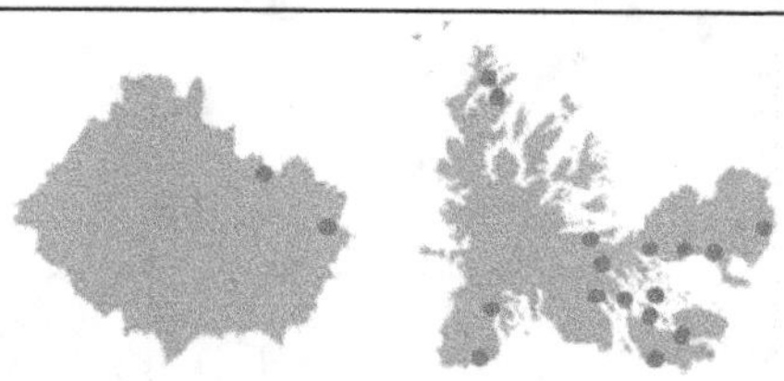

*Dendrodrilus rubidus**

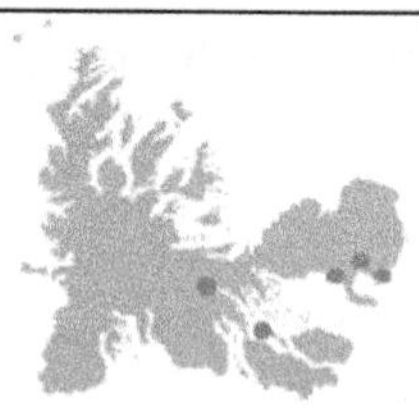

Eiseniella tetraedra

Mollusca Stylommatophora

Agriolimacidae

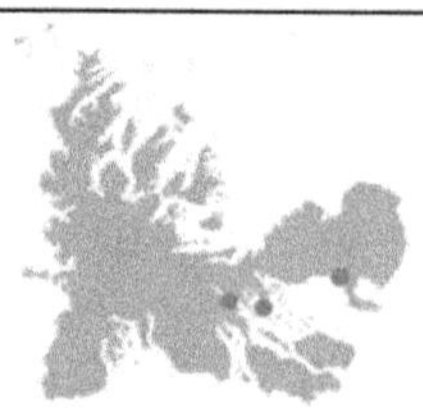

Deroceras invadens

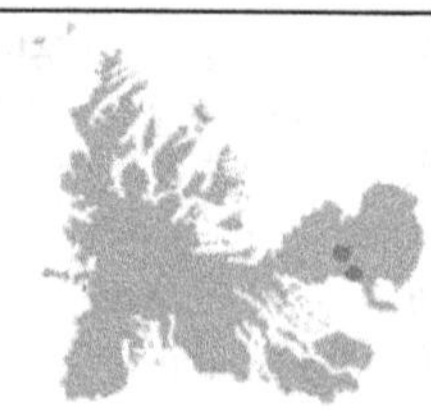

Deroceras reticulatum

Arachnida Araneae

Linyphiidae

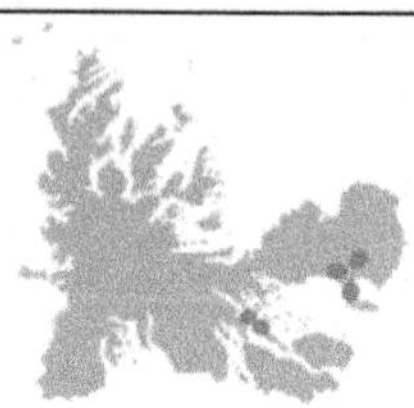

Tenuiphantes tenuis

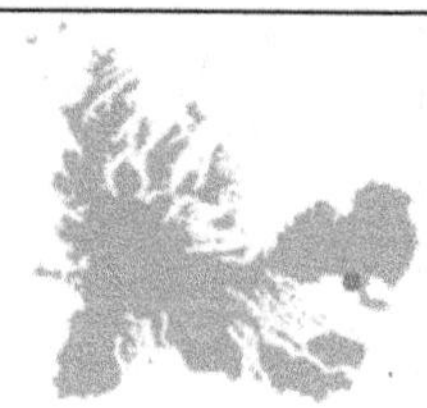

Lepthyphantes leprosus

Ostearius melanopygius

Agelinidae

Tegeneria domestica

Pholcidae

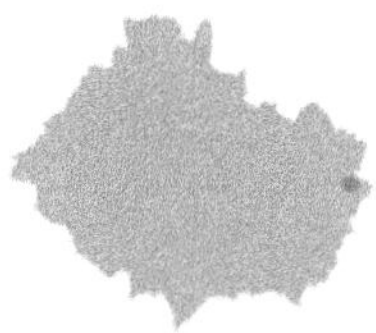

Pholcus phalangioides

Theridiidae

Steatoda grossa

Steatoda triabgulosa

Insecta Coleoptera

Latridiidae

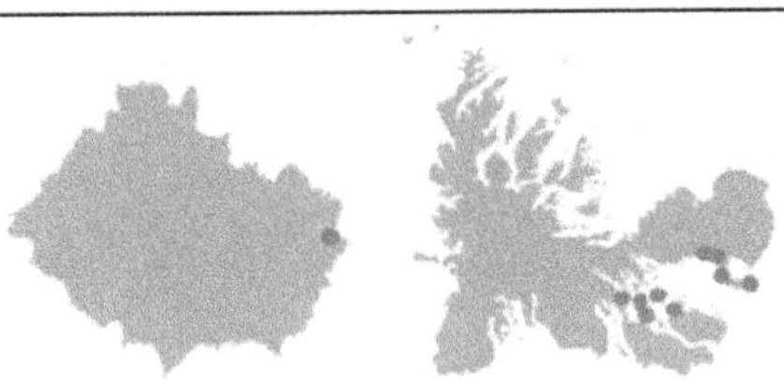

Cartodere nodifer

Latridius minutus

Ptinidae

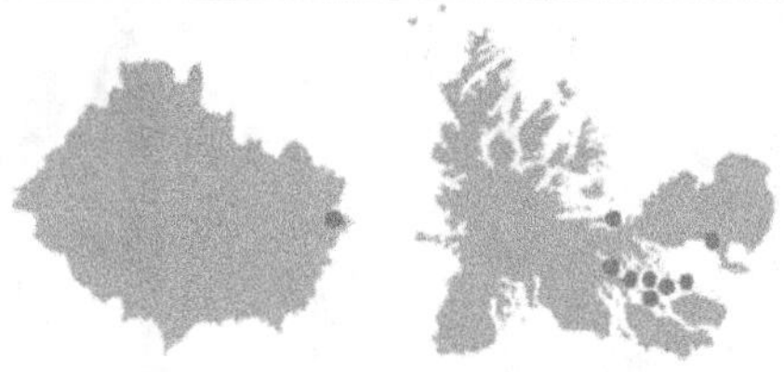

Ptinus tectus

Trechidae

*Merizodus soledadinus**

Insecta Diptera

Anthomyiidae

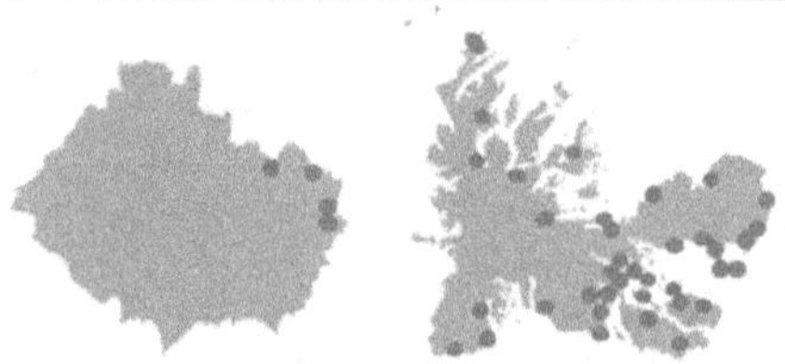

*Fucellia tergina**

Calliphoridae

*Calliphora vicina**

Carnidae

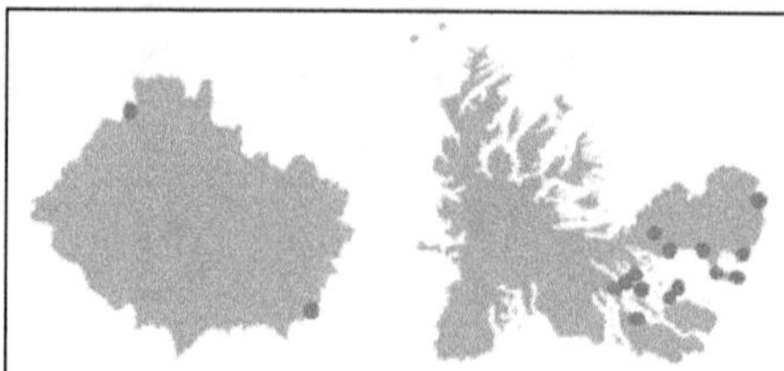

Hemeromya (?) sp.*

Chironomidae

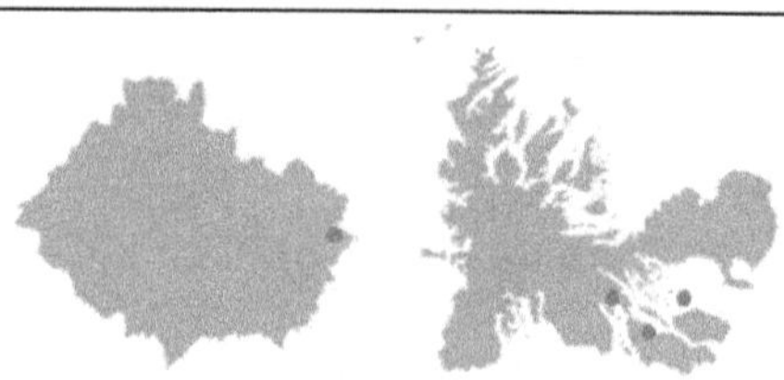

Bryophaenocladius sp.

Chironomidae

*Limnophyes minimus**

Smittia sp.*

Psychodidae

*Psychoda parthenogenetica**

Scatopsidae

Scatopse notata

Sciaridae

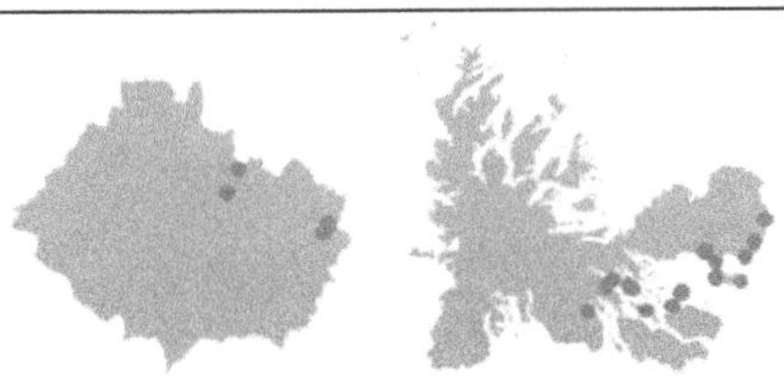

*Lycoriella sativae**

Sphaeroceridae

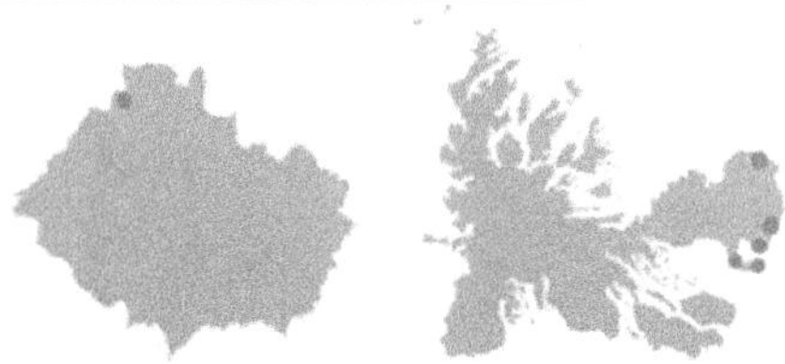

Leptocera curvinervis

Trichoceridae

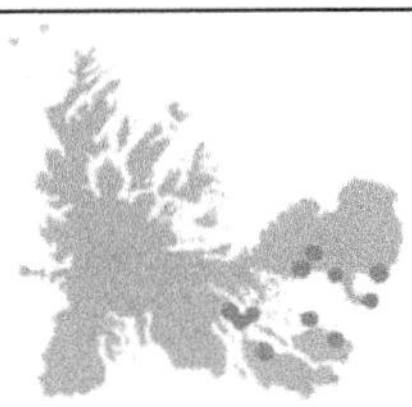

*Trichocera maculipennis**

Insecta Hemiptera

Aphididae

Aulacorthum solani

Macrosiphum euphorbiae

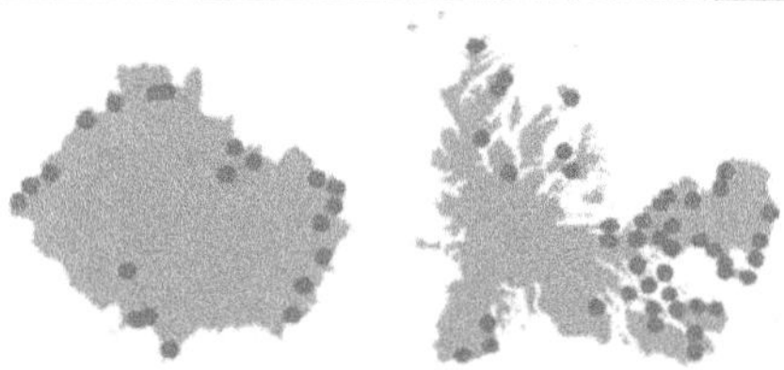

*Myzus ascalonicus**

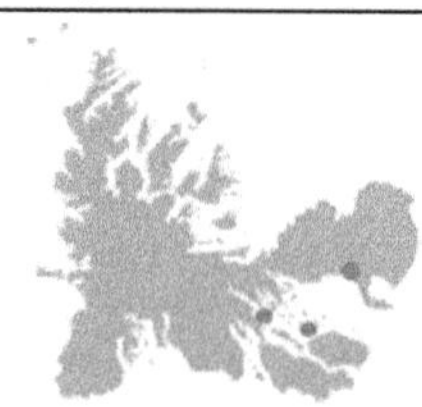

Myzus ornatus

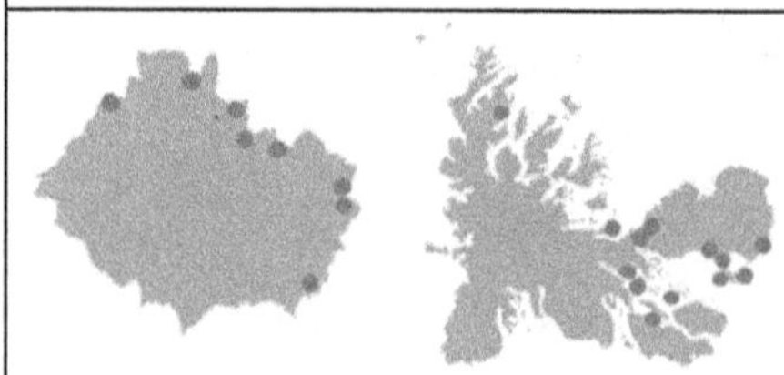

*Rhopalosiphum padi**

Insecta Psocoptera

Trogiidae

Lepinotus patruelis

Insecta Thysanoptera

Thripidae

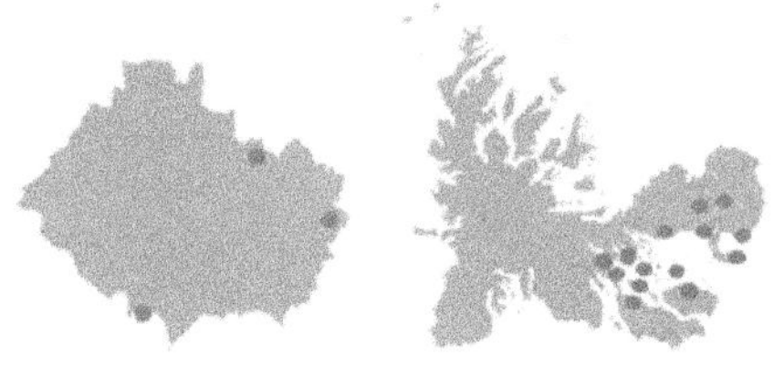

*Apterothrips apteris**

References

Allegrucci, G., Carchini, G., Convey, P., and Sbordoni, V. (2012). Evolutionary geographic relationships among orthocladine chironomid midges from maritime Antarctic and sub-Antarctic islands. *Biological Journal of the Linnean Society*, 106, 256–274.

Amice, G. (2003). Compte-rendu sur l'étude de l'hyménoptère parasitoïde *Kleidotoma icarus* (Quinlan, 1967), Île de la Possession. Paper, University of Rennes, France, 1, 38.

Anonymous (1879). Compte rendu des recherches sur la faune de l'Île de Kerguelen. *Annales des sciences naturelles*, 6, 1–30.

Arif, M., Ochoa-Corona, F.M., Opit, G.P., Li, Z., Kučerová, Z., Stejskal, V., and Yang, Q. (2012). PCR and isothermal-based molecular identification of the stored-product psocid pest *Lepinotus reticulatus* (Psocoptera: Trogiidae). *Journal of Stored Products Research*, 49, 184–188.

Badonnel, A. (1943). Psocoptères. *Faune de France*, 42, 1–164.

Badonnel, A. (1947). Croisière du Bougainville aux îles australes françaises. XIV. Psocoptères. *Mémoires du Muséum, Nouvelle Série*, XX, 25–30.

Badonnel, A. (1970). *Antarctopsocus daviesi*, nouvelle espèce des Îles Crozet (Psocoptera, Elipsocidae). *Bulletin du Muséum National d'Histoire Naturelle*, 42, 355–362.

Bailly, S. and Dreux, P. (1992). Caryotypes de *Neocanonopsis dreuxi* Hoffmann et *Christensenia Antarctica* Brinck, Curculionidae Ectemnorrhininae (Insectes, Coléoptères) de l'archipel Crozet. *Génétique, sélection et évolution*, 24, 289–303.

Barbut, J. and Voisin, J.F. (2014). Deux espèces de Noctuidae introduites dans l'archipel Crozet (Océan Indien Austral) (Lepidoptera). *L'Entomologiste*, 70, 121–122.

Barendse, J. and Chown, S.L. (2001). Abundance and seasonality of mid-altitude fellfield arthropods from Marion Island. *Polar Biology*, 24, 73–82.

Bartish, I.V., Aïnouche, A., Jia, D., Bergstrom, D., Chown, S.L., Winkworth, R.C., and Hennion, F. (2012). Phylogeny and colonization history of *Pringlea antiscorbutica* (Brassicaceae), an emblematic endemic from the South Indian Ocean Province. *Molecular Phylogenetics and Evolution*, 65, 748–756.

Basberg, B.L. and Headland, R.K. (2008). The 19th Century Antarctic sealing industry. Sources, data and economic significance. *SCAR/IASC IPY Open Science Conference*, St Petersburg, Norges Handelshoyskole, 1–24.

Beaucournu-Saguez, F. and Vernon, P. (1990). *Crozetia seguyi*, n. sp. (Diptera: Simuliidae) pour *Cnephia crozetensis* (Womersley, 1937) *sensu* Séguy (1940), espèce nouvelle des Îles Crozet. *Annales de la Société Entomologique de France*, 26, 405–409.

Bellés, X., Dreux, P., and Voisin, J.F. (1989). *Ptinus tectus* Boieldieu (Coleoptera, Ptinidae) aux Îles Kerguelen. *Bulletin de la Société Entomologique de France*, 93, 256.

Bellido, A. (1982). Les biocénoses du littoral rocheux aux Îles Kerguelen. *Colloque sur les écosystèmes subantarctiques, CNFRA (Comité National Français des Recherches Antarctiques)*, 51, 81–92.

Berland, L. (1947). Croisière du Bougainville aux îles australes françaises. XVI. Araignées. *Mémoires du Muséum, Nouvelle Série*, XX, 53–64.

Beron, P. (2018). Zoogeography of Arachnida. *Monographiae Biologicae*, 94, 1–937.

Blackman, R.L. and Eastop, V.F. (2000). *Aphids on the World's Crops: An Identification and Information Guide*, 2nd edition. John Wiley, Chichester.

Blakemore, R.J. (2008). Review of Southern Ocean, South Atlantic and sub-Antarctic Island earthworms updated from Lee (1994). In *A Series of Searchable Texts on Earthworm Biodiversity, Ecology and Systematics from Various Regions of the World*, 3rd edition, Kaneko, N. and Ito, M.T. (eds). CD Edition, Yokohama, Japan.

Bouché, M.B. (1972). Lombriciens de France. Ecologie et Systématique. *Annales de Zoologie*, 72, 671.

Bouché, M.B. (1982). Les Lombriciens (Oligochaeta) des terres australes françaises. *Colloque sur les écosystèmes subantarctiques, CNFRA (Comité National Français des Recherches Antarctiques)*, 51, 175–180.

Bouget, C. and Vincent, R. (2008). Les Latridiidae de France continentale et de Corse : mise à jour de la clé des genres et du catalogue des espèces (Coleoptera, Cucujoidea). *Bulletin de la Société Entomologique de France*, 113, 110–120.

Brake, I. (1997). Ecological studies on Carnidae (Diptera) on the bird island Mellum. *Studia Dipterologica*, 4, 201–210.

Brignoli, P.M. (1981). New or interesting Anapidae (Arachnida, Araneae). *Revue Suisse de Zoologie*, 88, 109–134.

Broadley, A., Kauschke, E., and Mohrig, W. (2018). Black fungus gnats (Diptera: Sciaridae) found in association with cultivated plants and mushrooms in Australia, with notes on cosmopolitan pest species and biosecurity interceptions. *Zootaxa*, 4415, 201–242.

Brundin, L. (1962). Insects of Macquarie Island. Diptera: Chironomidae. *Pacific Insects*, 4, 945–954.

Charrier, M., Marie, A., Guillaume, D., Bédouet, L., Le Lannic, J., Roiland, C., Berland, S., Pierre, J.S., Le Floch, M., Frenot, Y., and Lebouvier, M. (2013). Soil calcium availability influences shell ecophenotype formation in the Sub-Antarctic land snail, *Notodiscus hookeri*. *PLoS ONE*, 8, e84527.

Chauvin, G. and Vernon, P. (1982). Quelques données sur la biologie et la systématique des Lépidoptères Subantarctiques (Îles Crozet, Îles Kerguelen). *Colloque sur les écosystèmes subantarctiques, CNFRA (Comité National Français des Recherches Antarctiques)*, 51, 101–110.

Chauvin, G., Vannier, G., and Vernon, P. (1986). L'oeuf d'*Anatalanta aptera* Eaton (Diptera: Sphaeroceridae), morphologie et résistance à la dessiccation. *CNFRA (Comité National Francais des Recherches Antarctiques)*, 58, 129–136.

Chevrier, M. (1996). Introduction de deux espèces d'insectes aux Îles Kerguelen : processus de colonisation et exemples d'interactions. Thesis, University of Rennes 1, Rennes, France.

Chevrier, M., Vernon, P., and Frenot, Y. (1997). Potential effects of two alien insects on a sub-Antarctic wingless fly in the Kerguelen Islands. In *Antarctic Communities: Species, Structure and Survival*, Battaglia, B., Valencia, J., and Walton, D.W.H. (eds). Cambridge Univesity Press, Cambridge, 424–431.

Chown, S.L. (1989). Habitat use and diet as biogeographic indicators for sub-Antarctic Ectemnorhinini (Coleoptera: Curculionidae). *Antarctic Science*, 1, 23–30.

Chown, S.L. (1992). A preliminary analysis of weevil assemblages in the sub-Antarctic: Local and regional patterns. *Journal of Biogeography*, 19, 87–98.

Chown, S.L. (1993). Instar number and mass of *Palirhoeus eatoni* (C.O. Waterhouse) and *Bothrometopus randi* Jeannel (Coleoptera: Curculionidae) from sub-Antarctic Marion Island. *The Coleopterists Bulletin*, 47, 69–73.

Chown, S.L. (1994). Historical ecology of sub-Antarctic weevils (Coleoptera: Curculionidae): Patterns and processes on isolated islands. *Journal of Natural History*, 28, 411–433.

Chown, S.L. and Convey, P. (2016). Antarctic entomology. *Annual Review of Entomology*, 61, 119–137.

Chown, S.L. and Froneman, P.W. (eds) (2008). *The Prince Edward Islands, Land-Sea Interactions in a Changing Ecosystem.* Sun Press, Stellenbosch, South Africa.

Chown, S.L. and Klok, C.J. (2001a). First record of *Palirhoeus eatoni* (Coleoptera: Curculiondae) from sub-Antarctic Heard Island. *African Entomology*, 9, 193–194.

Chown, S.L. and Klok, C.J. (2001b). Habitat use, diet and body size of Heard Island weevils. *Polar Biology*, 24, 706–712.

Chown, S.L. and Kuschel, G. (1994). New *Bothrometopus* species from Possession Island, Crozet Archipelago, with nomenclatural amendments and a key to its weevil fauna (Coleoptera: Curculionidae: Brachycerinae). *African Entomology*, 2, 149–154.

Chown, S.L. and van Drimmelen, M. (1992). Water balance and osmoregulation in weevil larvae (Coloeoptera: Curculionidae: Brachycerinae) from three different habitats on sub-Antarctic Marion Island. *Polar Biology*, 12, 527–532.

Chown, S.L., McGeoch, M.A., and Marshall, D.J. (2002). Diversity and conservation of invertebrates on the sub-Antarctic Prince Edward Islands. *African Entomology*, 10, 67–82.

Chown, S.L., Klok, C.J., and McGeoch, M.A. (2004). Weather to go out: Activity of *Bothrometopus brevis* (Curculionidae) at Heard Island. *Polar Biology*, 27, 217–221.

Colless, D.H. (1970). A new species of Macrocera from Crozet Islands (Diptera, Mycetophilidae). *Proceedings of the Royal Entomological Society of London. Series B, Taxonomy*, 39, 27–28.

Common, L.F.B. (1970). Lepidoptera: Yponomeutidae of Heard Island. *Pacific Insects Monograph*, 23, 229–233.

Convey, P. (2005). Recent lepidopteran records from sub-Antarctic South Georgia. *Polar Biology*, 28, 108–110.

Convey, P. (2011). Antarctic terrestrial biodiversity in a changing world. *Polar Biology*, 34, 1629–1641.

Convey, P., Key, R.S., and Key, R.J.D. (2010). The establishment of a new ecological guild of pollinating insects on sub-Antarctic South Georgia. *Antarctic Science*, 22, 508–512.

Coulson, S.J., Hodkinson, I.D., Webb, N.R., Mikkola, K., Harrison, J.A., and Pedgley, D.E. (2002). Aerial colonization of high Arctic islands by invertebrates: The diamondback moth *Plutella xylostella* (Lepidoptera: Yponomeutidae) as a potential indicator species. *Diversity and Distributions*, 8, 327–334.

Crafford, J.E. (1984). Life cycle and kelp consumption of *Paractora dreuxi mirabilis* (Diptera: Helcomyzidae): A primary decomposer of stranded kelp on Marion Island. *South African Journal of Antarctic Research*, 14, 18–22.

Crafford, J.E. and Chown, S.L. (1987). *Plutella xylostella* L (Lepidoptera, Plutellidae) on Marion Island. *Journal of the Entomological Society of Southern Africa*, 50, 259–260.

Crafford, J.E. and Scholtz, C.H. (1986). Impact of *Embryonopsis halticella* Eaton larvae (Lepidoptera: Yponomeutidae) feeding in Marion Island tussock grassland. *Polar Biology*, 6, 191–196.

Crafford, J.E., Scholtz, C.H., and Chown, S.L. (1986). The insects of sub-Antarctic Marion and Prince Edward Islands; with a bibliography of entomology of the Kerguelen Biogeographical Province. *South African Journal of Antarctic Research*, 16, 42–84.

Craig, D.A., Currie, D.C., and Vernon, P. (2003). *Crozetia* Davies (Diptera: Simulidae): Redescription of *Cr. crozetensis*, *Cr. seguyi*, number of larval instars, phylogenetic relationships and historical biogeography. *Zootaxa*, 259, 1–39.

Dahl, C. (1970). Diptera: Trichoceridae of Kerguelen Island. *Pacific Insects Monograph*, 23, 274–275.

Darlington Jr., P.J. (1970). Coleoptera: Carabidae of South Georgia. *Pacific Insects Monograph*, 23, 234.

Dartnall, H.J.G. (2017). The freshwater fauna of the south polar region: A 140-year review. *Papers and Proceedings of the Royal Society of Tasmania*, 151, 19–58.

Davies, L. (1972). Two *Amblystogenium* species (Col. Carabidae) co-existing on the sub-Antarctic Possession Island, Crozet Islands. *Entomologia Scandinaviae*, 3, 275–286.

Davies, L. (1973). Observations on the distribution of surface-living land arthropods on the sub-Antarctic Île de la Possession, Îles Crozet. *Journal of Natural History*, 7, 241–253.

Davies, L. (1974). Evolution of larval head-fans in Simuliidae (Diptera) as inferred from the structure and biology of *Crozetia crozetensis* (Womersley) compared to other genera. *Zoological Journal of the Linnean Society*, 55, 193–224.

Davies, L. (1982). Ecology of two *Amblystogenium* species (Carabidae) on Île de la Possession, Îles Crozet. *Colloque sur les écosystèmes subantarctiques, CNFRA (Comité National Français des Recherches Antarctiques)*, 51, 167–173.

Davies, L. (1987). Long adult life, low reproduction and competition in two sub-Antarctic carabid beetles. *Ecological Entomology*, 12, 149–162.

Davies, L. (2006). Lifetime reproductive output of *Calliphora vicina* and *Lucilia sericata* in outdoor caged and field populations; flight vs. egg production? *Medical and Veterinary Entomology*, 20, 453–458.

Davies, L. and Greene, S.W. (1976). Notes sur la végétation de l'Île de la Possession (Archipel Crozet). *CNFRA (Comité National Français des Recherches Antarctiques)*, 41, 1–20.

Davies, K.F., Greenslade, P., and Melbourne, B.A. (1997). The invertebrates of sub-Antarctic Bishop Island. *Polar Biology*, 17, 455–458.

Davies, L., Bouvet, S., and Vernon, P. (2007). All-year reproduction and possible thermal melanism in *Amblystogenium pacificum* (Coleoptera: Carabidae) on the sub-Antarctic Île de la Possession (Iles Crozet). *Polar Biology*, 30, 253–260.

De Blignières, F.X. and Tréhen, P. (1982). Rythme circadien et phototaxie chez *Anatalanta aptera* (Eaton), Diptère Sphaeroceridae des Îles Kerguelen, approche de la signification biologique. *Colloque sur les écosystèmes subantarctiques, CNFRA (Comité National Francais des Recherches Antarctiques)*, 51, 157–165.

De Queiroz, A. (2005). The resurrection of oceanic dispersal in historical biogeography. *Trends in Ecology & Evolution*, 20, 68–73.

Delépine, G. (1973). Toponymie des terres australes. Report, Territoire des terres australes et antarctiques françaises et Commission territoriale de toponymie, Paris.

Delépine, G. (1995). *Les Îles australes françaises*. Editions Ouest-France, Rennes, France.

Delettre, Y. (1978). Biologie et écologie de *Limnophyes pusillus* Eaton, 1875 (Diptera, Chironomidae) aux Îles Kerguelen. 1 – Présentation générale et étude des populations larvaires. *Revue d'écologie et de biologie du sol,* 15, 475–486.

Delettre, Y. (1983). Sur le régime alimentaire des larves de deux Chironomides (Diptera) des lands armoricaines. *New Trends in Soil Biology. Proceedings of the VIII International Colloquium of Soil Biology*, Louvain-la-Neuve. Dieu-Brichard, Ottignies, Belgium, 616–617.

Delettre, Y. and Cancela Da Fonseca, J.P. (1979). Biologie et écologie de *Limnophyes pusillus* Eaton, 1875 (Diptera, Chironomidae) aux Îles Kerguelen. 2 – Etude des populations imaginales et discussion. *Revue d'écologie et de biologie du sol,* 16, 355–372.

Delettre, Y.R., Frenot, Y., Vernon, P., and Chown, S.L. (2003). First record of *Telmatogeton* sp. (Diptera: Chironomidae) at Heard Island. *Polar Biology*, 26, 423–426.

Delgado, J.A. and Palma, R.L. (1998). Larval stages of three *Meropathus* species (Coleoptera: Hydraenidae: Ochthebiinae) from New Zealand. *New Zealand Journal of Zoology*, 25, 409–419.

Doncaster, J.P. and Fres, M.A. (1946). The shalott aphis, *Myzus ascalonicus* sp. n. (Hemiptera, Aphididae). *Proceedings of the Royal Society of London. Series B: Biological Sciences*, 15, 3–4.

Dreux, P. (1965a). Faune entomologique de l'Île aux Cochons. *TAAF (Terres Australes et Antarctiques Françaises)*, 31, 57–69.

Dreux, P. (1965b). Faune entomologique de l'Île de la Possession (archipel Crozet). *TAAF (Terres Australes et Antarctiques Françaises)*, 30, 58–73.

Dreux, P. (1966). La faune entomologique de l'archipel Crozet et son endémisme. *CNFRA (Comité National Francais des Recherches Antarctiques)*, 15, 35–55.

Dreux, P. (1970). Contribution à l'étude des arthropodes terrestres et d'eau douce des Terres Australes Françaises. *TAAF (Terres Australes et Antarctiques Françaises)*, 50, 41–44.

Dreux, P. (1971a). La faune entomologique des Îles Marion et du Prince Edouard. In *Marion and Prince Edward Islands*, Van Zinderen Bakker Sr, E.M., Winterbottom, J.M., and Dyer, R.A. (eds). Balkema, A.A., Cape Town, 335–343.

Dreux, P. (1971b). Premières données sur la faune entomologique de l'Île de l'Est (Archipel Crozet). *TAAF (Terres Australes et Antarctiques Françaises)*, 45–54.

Dreux, P. (1978). La faune entomologique de l'Île Heard. *Bulletin de la Société Entomologique de France*, 83, 108–115.

Dreux, P. and Voisin, J.F. (1969). Note sur la faune des Coléoptères de l'Île de la Possession (Archipel Crozet). *Bulletin de la Société Entomologique de France*, 74, 55–56.

Dreux, P. and Voisin, J.F. (1978). Note sur le genre *Dusmoecetes* et description de *D. tamaris* n. sp. (Col., Curculionidae Ectemnorhininae). *Bulletin de la Société Entomologique de France*, 83, 29–36.

Dreux, P. and Voisin, J.F. (1983). Sur les genres *Christensenia* Brinck et *Neocanonopsis* Hoffman (Coléoptères Curculionidae Ectemnorhininae). *L'Entomologiste*, 39, 209–214.

Dreux, P. and Voisin, J.F. (1984). Description de *Bothrometopus derelictorum*, n. sp. et notes sur le genre *Bothrometopus* Jeannel, 1940. (Coleoptera, Curculionidae, Ectemnorhininae). *Revue française d'entomologie (N.S.)*, 6, 33–38.

Dreux, P. and Voisin, J.F. (1985). Notes sur le genre *Xanium* Enderlein (Coleoptera, Curdulionidae). *Nouvelle Revue d'Entomologie*, 2, 431–436.

Dreux, P. and Voisin, J.F. (1986a). Note sur les genres *Mesembriorhinus* Jeannel et *Palirhoeus* Kuschel (Coleoptera, Curculionidae). *Nouvelle Revue d'Entomologie*, 3, 257–261.

Dreux, P. and Voisin, J.F. (1986b). Révision du genre *Dusmoecetes* Jeannel 1940, et description de deux espèces et d'une sous-espèce nouvelles (Col. Curculionidae Ectemnorhininae). *Bulletin de la Société Entomologique de France*, 91, 27–40.

Dreux, P. and Voisin, J.F. (1987). Notes sur le genre *Antarctonesiotes* Jeannel et sur le genre *Disker* nov., et description d'*Antarctonesiotes villiersi* n. sp. (Coleoptera, Curculionidae). *Nouvelle Revue d'Entomologie*, 4, 313–321.

Dreux, P. and Voisin, J.F. (1988). Sur le genre *Ectemnorrhinus* G.R. Waterhouse 1853 (Coleoptera, Curculionidae). *Nouvelle Revue d'Entomologie*, 5, 375–382.

Dreux, P. and Voisin, J.F. (1992). Entomological fauna of the Île des Pingouins (Crozet archipelago). *Bulletin de la Société Entomologique de France*, 97, 453–464.

Dreux, P. and Voisin, J.F. (1993). Faune entomologique de l'île des Pingouins (archipel Crozet). *Bulletin de la Société Entomologique de France*, 97, 453–464.

Dreux, P., Frenot, Y., Jouventin, P., Massé, L., and Verdier, O. (1988). First scientific expedition to Ile des Pingouins, Archipel Crozet. *Polar Record*, 24, 235–238.

Dreux, P., Galiana, D., and Voisin, J.F. (1992). Acclimatation de *Merizodus soledadinus* Guérin dans l'archipel de Kerguelen (Coleptera, Trechidae). *Bulletin de la Société Entomologique de France*, 97, 219–221.

Duckhouse, D.A. (1970). Diptera: Psychodidae of South Georgia. *Pacific Insects Monograph*, 23, 283.

Duffy, G.A., Coetzee, B.W.T., Latombe, G., Akerman, A.H., McGeogh, M.A., and Chown, S.L. (2017). Barriers to globally invasive species are weakening across the Antarctic. *Diversity and Distributions*, 23, 982–996.

Eastop, V.F. (1962). Insects of Macquarie Island. Hemiptera: Homoptera: Aphididae. *Pacific Insects*, 4, 937–938.

Eastop, V.F. (1970). Hemiptera: Homoptera: Aphididae of South Georgia. *Pacific Insects Monograph*, 23, 227.

Eaton, A.E. (1879). The collections form Kerguelen Island: Introductory notes. *Philosophical Transactions of the Royal Society of London*, 168, 1–8.

Enderlein, G. (1903). Die Landarthropoden der von der Tiefsee-Expéedition besuchten antarktischen Inseln. In *Wissenchaftliche Ergebnisse der Deutschen Tiefsee-Expedition auf dem Dampfer "Valdivia", 1898–1899*, Chun, C. (ed.). Jena, Germany, 218–290.

Enderlein, G. (1905). Beitrag zur Kenntnis der antarktischen Landarthropoden. Pringleophaga, eine neue Schmetterlingsgattung aus dem antarktischen Gebiet. *Zoologischer Anzeiger – A Journal of Comparative Zoology*, 29, 119–125.

Enderlein, G. (1908). Die Insekten des Antarktischen Gebietes. In *Deutsche Südpolar-Expedition 1901–1903*, Von Drygalski, E. (ed.). Druck und Verlag Georg Reimer, Berlin, 365–528.

Endler, J.A. (1982). Alternative hypotheses in biogeography: Introduction and synopsis of the symposium. *American Zoologist*, 22, 349–354.

Evenhuis, N.L. (1989). Appendix I. Diptera of Antarctica and sub-Antarctic islands. In *Catalog of Oceanic and Australian Diptera*, Evenhuis, N.L. (ed.). Bishop Museum Press, Leiden and EJ Brill, Honolulu, Hawaii, 797–804.

Fage, L. (1940). Croisière du Bougainville aux îles australes françaises. VII. Opiliones. *Mémoires du Muséum National d'Histoire Naturelle*, 14, 283–284.

Falk-Petersen, J., Bøhn, T., and Sandlund, O. (2006). On the numerous concepts in invasion biology. *Biological Invasions*, 8, 1409–1424.

Forel, A. (1908). Fourmis de Ceylan et d'Egypte récoltées par le Prof. E. Bugnon. Lasius carniolicus. Fourmis de Kerguelen. Pseudandrie? Strongylognathus testaceus. *Bulletin de la Société Vaudoise des Sciences Naturelles*, 44(162), 1–23.

Forel, A. (1914). Formicides d'Afrique et d'Amérique nouveaux ou peu connus. *Bulletin de la Société Vaudoise de Sciences Naturelles*, 50(184), 211–364.

Frank, J.H. and Ahn, K.-J. (2011). Coastal Staphylinidae (Coleoptera): A worldwide checklist, biogeography and natural history. *ZooKeys*, 107, 1–98.

Frenot, Y. (1985). Etude de l'introduction accidentelle de *Dendrobaena rubida tenuis* (Oligochaeta, Lumbricidae) à l'Île de la Possession. *Bulletin d'Ecologie*, 16, 47–53.

Frenot, Y. (1986a). Interactions entre la faune lombricienne et les systèmes édaphiques d'une île subantarctique : Île de la Possession, Archipel Crozet. Thesis, University of Rennes 1, Rennes, France.

Frenot, Y. (1986b). Organisation des communautés de vers de terre de l'île de la Possession, Archipel Crozet. *CNFRA (Comité National Francais des Recherches Antarctiques)*, 58, 215–230.

Frenot, Y., Tréhen, P., and Vernon, P. (1987). Organisation comparée des peuplements de diptères et de lombriciens en milieu insulaire subantarctique : exemples de l'Île de la Possession, de l'Île aux Cochons (Archipel Crozet) et de l'Île de Croÿ (Archipel Kerguelen). *Bulletin de la Société Zoologique de France*, 112, 484–494.

Frenot, Y., Chown, S.L., Whinam, J., Selkirk, P.M., Convey, P., Skotnicki, M., and Bergstrom, D.M. (2005). Biological invasions in the Antarctic: Extent, impacts and implications. *Biological Reviews*, 80, 45–72.

Frenot, Y., Convey, P., Lebouvier, M., Chown, S.L., Whinam, J., Selkirk, P.M., Skotnicki, M., and Bergstrom, D.M. (2008). Antarctic and sub-Antarctic biological invasions: Sources, extents, impacts and implications. In *Non-native Species in the Antarctic*, Rogan-Finnemore, M. (ed.). Gateway Antarctica Special Publication Series, Christchurch, New Zealand.

Gadea, A., Le Pogam, P., Biver, G., Boustie, J., Le Lamer, A.C., Le Dévéhat, F., and Charrier, M. (2017). Which specialized metabolites does the native sub-Antarctic gastropod *Notodiskus hookeri* extract from the consumption of the lichens *Usnea taylorii* and *Pseudocyphellaria crocata*? *Molecules and Cells*, 22, 425.

Gaston, K.J., Chown, S.L., and Bradley, P. (2002). Invertebrate diversity and endemism at Gough Island and threats from introduced species. Final project report, Darwin Initiative for the Survival of Species.

Gaston, K.J., Jones, A.G., Hänel, C., and Chown, S.L. (2003). Rates of species introduction to a remote oceanic island. *Proceedings of the Royal Society of London. Series B: Biological Sciences*, 270(1519), 1091–1098.

Giret, A., Weis, D., Grégoire, M., Mattielli, N., Moine, B., Michon, G., Scoates, J., Tourpin, S., Delpech, G., Gerbé, M.C., Doucet, S., Ethien, R., and Cottin, J.Y. (2003a). L'Archipel de Kerguelen : les plus vieilles îles dans le plus jeune océan. *Géologues*, 137, 23–39.

Giret, A., Weis, D., Zhou, X., Cottin, J.Y., and Tourpin, S. (2003b). Les Îles Crozet. *Géologues*, 137, 15–23.

Green, K. and Mound, L. (1994). An extension to the insect fauna of Heard Island. *Polar Record*, 30(173), 131–132.

Greenslade, P. (2006). *The Invertebrates of Macquarie Island.* Australian Antarctic Division, Kingston, Tasmania.

Greenslade, P., Farrow, R.A., and Smith, J.M.B. (1999). Long distance migration of insects to a sub-Antarctic island. *Journal of Biogeography*, 26, 1161–1167.

Greenslade, P., Vernon, P., and Smith, D. (2012). Ecology of Heard Island Diptera. *Polar Biology*, 35, 841–850.

Gressitt, J.L. (1964). Insects of Campbell Island. Summary. *Pacific Insects Monograph*, 7, 531–600.

Gressitt, J.L. (1970). Sub-Antarctic entomology and biogeography. *Pacific Insects Monograph*, 23, 295–374.

Greve, M., Gremmen, N.J.M., Gaston, K.J., and Chown, S.L. (2005). Nestedness of Southern Ocean island biotas: Ecological perspectives on a biogeographical conundrum. *Journal of Biogeography*, 32, 155–168.

Greve, M., Mathakutha, R., Steyn, C., and Chown, S.L. (2017). Terrestrial invasions of sub-Antarctic Marion and Prince Edwards Islands. *Bothalia, African Biodiversity and Conservation*, 47, a2143.

Hagen, H.A. (1876). *Pseudo-Neuroptera*. Smithsonian Institution, Washington DC.

Hänel, C. (1999). The distribution and abundance of macro-invertebrates in the major vegetation communities of Marion Island and the impact of alien species. Study, Msc. University of Pretoria, Pretoria, South Africa.

Hänel, C. (2007). Terrestrial invertebrates. In *Field Guide to the Animals and Plants of Tristan da Cunha and Gough Island*, Ryan, P.G. (ed.). CTP Book, Parow, South Africa.

Hänel, C. and Chown, S.L. (1998). The impact of a small, alien invertebrate on a sub-Antarctic terrestrial ecosystem: *Limnophyes minimus* (Diptera, Chironomidae) at Marion Island. *Polar Biology*, 20, 99–106.

Hänel, C., Chown, S.L., and Davies, L. (1998). Records of alien insect species from sub-Antarctic Marion and South Georgia islands. *African Entomology*, 6, 366–369.

Harrison, R.A. (1970). Diptera: Acalypterates of South Georgia, Heard and Kerguelen. *Pacific Insects Monograph*, 23, 285–289.

Haupt, T.M., Crafford, J.E., and Chown, S.L. (2014). Solving the puzzle of *Pringleophaga* – threatened, keystone detritivores in the sub-Antarctic. *Insect Conservation and Diversity*, 7, 308–313.

Headland, R.K. (1989). *Chronological List of Antarctic Expeditions and Related Historical Events*. Cambridge University Press, Cambridge.

Hickman, V.V. (1939). Opiliones and Araneae. *Reports of British and New Zealand Antarctic Research Expedition, 1929–1931, Series B (Zoology and Botany)*, 4, 159–187.

Houghton, M., Terauds, A., Merritt, D., Driessen, M., and Shaw, J. (2019). The impacts of non-native species on the invertebrates of Southern Ocean Islands. *Journal of Insect Conservation*, 23, 435–452.

Hughes, K.A. and Convey, P. (2012). Determining the native/non-native status of newly discovered terrestrial and freshwater species in Antarctica – Current knowledge, methodology and management action. *Journal of Environmental Management*, 93, 52–66.

Hughes, K.A. and Convey, P. (2014). Alien invasions in Antarctica: Is anyone liable? *Polar Research*, 33, 22103.

Hughes, K.A., Lee, J.E., Tsujimoto, M., Imura, S., Bergstrom, D.M., Ware, C., Lebouvier, M., Huiskes, A.H.L., Gremmen, N.J.M., Frenot, Y., Bridge, P.D., and Chown, S.L. (2011). Food for thought: Risks of non-native species transfer to the Antarctic region with fresh produce. *Biological Conservation*, 144, 1682–1689.

Huiskes, A.H.L., Gremmen, N.J.M., Bergstrom, D.M., Frenot, Y., Hughes, K.A., Imura, S., Kiefer, K., Lebouvier, M., Lee, J.E., Tsujimoto, M., Ware, C., Van De Vijver, B., and Chown, S.L. (2014). Aliens in Antarctica: Assessing transfer of plant propagules by human visitors to reduce invasion risk. *Biological Conservation*, 171, 278–284.

Huiskes, A.H.L., Lebouvier, M., Molina Montenegro, M.A., Pertierra, L.R., and Tsujimoto, M. (2018). Pathways for the introduction of terrestrial non-native species. *Antarctic Environments Portal*. Available at: https://www.environments.aq/.

Hullé, M. (2012). *Myzus ascalonicus*, an aphid recently introduced to sub-Antarctic islands, prefers native to exotic host-plants. *Environmental Entomology*, 41, 1398–1404.

Hullé, M., Pannetier, D., Maurice, D., Courmont, L., Chaillon, C., Chaillon, P.-E., Saccone, P., Hébert, C., Gracia, M., Buffin, J., Simon, J.-C., and Frenot, Y. (2003a). Aphids from Kerguelen and Crozet Islands, sub-Antarctic. In *Antarctic Biology in a Global Context*, Huiskes, A.H.L., Gieskes, W.W.C., Rozema, J., Schorno, R.M.L., van der Vies, S.M., and Wolff, W.J. (eds). Backhuys Publishers, Leiden.

Hullé, M., Pannetier, D., Simon, J.-C., Vernon, P., and Frenot, Y. (2003b). Aphids of sub-Antarctic Îles Crozet and Kerguelen: Species diversity, host range and spatial distribution. *Antarctic Science*, 15, 203–209.

Hullé, M., Turpeau, E., Hudaverdian, S., Chaubet, B., Outreman, Y., and Lebouvier, M. (2010). Aphids and associated natural enemies on Ile Amsterdam and Ile Saint-Paul, Southern Indian Ocean. *Antarctic Science*, 22, 379–385.

Hullé, M., Buchard, C., Georges, R., and Vernon, P. (2018). Guide d'identification des invertébrés des îles sub-Antarctiques Kerguelen et Crozet. Report, University of Rennes 1, Rennes, France.

Hutchinson, J.M.C., Reise, H., and Robinson, D.G. (2014). A biography of an invasive terrestrial slug: The spread, distribution and habitat of *Deroceras invadens*. *NeoBiota*, 23, 17–64.

Jarrige, J. and Voisin, J.F. (1968). Note sur *Pseudeuplectus antarcticus* Enderlein. *Bulletin de la Société Entomologique de France*, 73, 59–61.

Jeannel, R. (1940). Croisière du Bougainville. III–Coléoptères. *Mémoires du Muséum National d'Histoire Naturelle*, XIV, 63–201.

Jeannel, R. (1953a). Sur la faune entomologique de l'île Marion. *Revue Française d'Entomologie*, 20, 161–167.

Jeannel, R. (1953b). Sur un Psélaphide des îles Crozet. *Revue Française d'Entomologie*, 20, 168–173.

Jeannel, R. (1965). Biogéographie des terres australes de l'océan Indien. *Revue Française d'Entomologie*, 31, 319–422.

Joly, Y., Frenot, Y., and Vernon, P. (1987). Environmental modifications of a sub-Antarctic peat-bog by the Wandering Albatross (*Diomedea exulans*): A preliminary study. *Polar Biology*, 8, 61–72.

Jones, A.G., Chown, S.L., and Gaston, K.J. (2002). Terrestrial invertebrates of Gough Island: An assemblage under threat? *African Entomology*, 10, 83–91.

Keulartz, J. and Van Der Weele, C. (2009). Between nativism and cosmopolitism: Framing and reframing in invasion biology. In *New Visions of Nature*, Drenthen, M., Keulartz, J., and Proctor, J. (eds). Springer, The Netherlands.

Kidder, J.H. (1876). Contributions to the natural history of Kerguelen Island. *Bulletin of the United States National Museum*, 3(II), 124.

Kuschel, G. (1970). Coleoptera: Curculionidae of Heard Island. *Pacific Insects Monograph*, 23, 255–260.

Kuschel, G. and Chown, S.L. (1995). Phylogeny and systematics of the *Ectemnorhinus*-group of genera (Insecta: Coleoptera). *Invertebrate Taxonomy*, 9, 841–863.

Laparie, M. (2011). Invasive success of two alien insects at the Kerguelen Islands: Role of morphological and ecophysiological adjustments to the novel environmental conditions. Thesis, University of Rennes 1, Rennes, France.

Laparie, M., Lebouvier, M., Lalouette, L., and Renault, D. (2010). Variation of morphometric traits in populations of an invasive carabid predator (*Merizodus soledadinus*) within a sub-Antarctic island. *Biological Invasions*, 12, 3405–3417.

Laparie, M., Bical, R., Larvor, V., Vernon, P., Frenot, Y., and Renault, D. (2012). Habitat phenotyping of two sub-Antarctic flies by metabolic fingerprinting: Evidence for a species outside its home. *Comparative Biochemistry and Physiology A*, 162, 406–412.

Lebouvier, M., Laparie, M., Hullé, M., Marais, A., Cozic, Y., Lalouette, L., Vernon, P., Candresse, T., Frenot, Y., and Renault, D. (2011). The significance of the sub-Antarctic Kerguelen Islands for the assessment of the vulnerability of native communities to climate change, alien insect invasions and plant viruses. *Biological Invasions*, 13, 1195–1208.

Ledoux, J.C. (1991). Araignées des îles subantarctiques françaises (Crozet et Kerguelen). *Revue Arachnologique*, 9, 119–164.

Lee, K.E. (1968). *Oligochaeta from Sub-Antarctic Islands*. Mawson Institute for Antarctic Research, The University of Adelaide, Adelaide.

Lee, J.E. and Chown, S.L. (2016). Range expansion and increasing impact of the introduced wasp *Aphidius matricariae* Haliday on sub-Antarctic Marion Island. *Biological Invasions*, 18, 1235–1246.

Lehnebach, C.A., Winkworth, R.C., Becker, M., Lockhart, P.J., and Hennion, F. (2017). Around the pole: Evolution of sub-Antarctic *Ranunculus*. *Journal of Biogeography*, 44, 875–886.

Le Roux, V., Chapuis, J.L., Frenot, Y., and Vernon, P. (2002). Diet of the house mouse (*Mus musculus*) on Guillou Island, Kerguelen archipelago, sub-Antarctic. *Polar Biology*, 25, 49–57.

Leuchtmann, A. and Michelsen, V. (2016). Biology and evolution of the *Epichloë*-associated *Botanophila* species found in Europe (Diptera: Anthomyiidae). *Insect Systematics & Evolution*, 47, 1–14.

Lienhard, C. (1998). Psocoptères Euro-Méditerranéens. *Faune de France. Fédération Française des Sociétés de Sciences Naturelles*, 83, 550.

Luff, M.L. (1972). A new species of *Amblystogenium* Enderlein (Coleoptera: Carabidae) from the Crozet Islands, with comparative notes with *A. pacificum* (Putzeys). *Journal of Entomology (B)*, 41, 53–58.

MacArthur, R.H. and Wilson, E.O. (1967). *The Theory of Island Biogeography*. Princetown University Press, New Jersey.

Mammola, S. (2019). Finding answers in the dark: Caves as models in ecology fifty years after Poulson and White. *Ecography*, 42, 1331–1351.

Marris, J.W.M. (2000). The beetle (Coleoptera) fauna of the Antipodes Islands, with comments on the impact of mice; and an annotated checklist of the insect and arachnid fauna. *Journal of the Royal Society of New Zealand*, 30, 169–195.

Mathis, W.N. (2011). World catalog and conspectus on the family Helcomyzidae (Diptera: Schizophora). *Myia*, 12, 267–280.

Matile, L. (1975). Un nouveau cas de brachyptérisme ches les Diptères Mycetophilidae : la femelle de *Macrocera crozetensis* Colless. *Bulletin de la Société Entomologique de France*, 80, 2–6.

McGeogh, M.A., Shaw, J.D., Terauds, A., Lee, J.E., and Chown, S.L. (2015). Monitoring biological invasion across the broader Antarctic: A baseline and indicator framework. *Global Environmental Change*, 32, 108–125.

Mercer, R.D., Chown, S.L., and Marshall, D.J. (2000). Mite and insect zonation on a Marion Island rocky shore: A quantitative approach. *Polar Biology*, 23, 775–784.

Michaux, B. and Leschen, R.a.B. (2005). East meets west: Biogeology of the Campbell Plateau. *Biological Journal of the Linnean Society*, 86, 95–105.

Munari, L. (2007). Studies on the Canacidae (Diptera), subfamily Apetaeninae. I. *Apetaenus enderleini*, nomen novum for *Listriomastax litorea* Enderlein, 1909, with remarks on the chaetotaxy, morphology, and habitats of the Apetaeninae from the Kerguelen Biogeographical Province. *Zootaxa*, (1542), 21–34.

Munari, L. (2008). Studies on the Canacidae (Diptera), subfamily Apetaeninae. II. A review of the world subgenera of *Apetaenus* Eaton, with a special reference of the Australian and New Zealand species. *Zootaxa*, 1692, 26–42.

Nakahara, S. (1988). A new synonym and revised status in *Apterothips* (Thysanoptera: Thripidae). *Proceedings of the Entomological Society of Washington*, 90, 508–509.

New, T.R. (1974). Psocoptera. *Handbooks for the Identification of British Insects*, 1, 104.

Orlov, I., Solodovnikov, A., and Leschen, R.a.B. (2019). Systematics and biogeography of the sub-Antarctic Leptusa (Coleoptera: Staphylinidae: Aleocharinae: Homalotini). *Systematics and Biodiversity*, 17, 60–79.

Ouisse, T., Laparie, M., Lebouvier, M., and Renault, D. (2017). New insights into the ecology of *Merizodus soledadinus*, a predatory carabid beetle invading the sub-Antarctic Kerguelen Islands. *Polar Biology*, 40, 2201–2209.

Palmer, J.M. (1974). Arthropoda of the sub-Antarctic islands of New Zealand (2) Hemiptera: Aphididae. *Journal of the Royal Society of New Zealand*, 4, 303–306.

Papierok, B., Dedryver, C.A., and Hullé, M. (2016). First records of aphid-pathogenic Entomophthorales in the sub-Antarctic archipelagos of Crozet and Kerguelen. *Polar Research*, 35, 28765.

Papp, L. (1983). Taxonomic notes on some flies of the Crozet Islands (Diptera). *Folia Entomologica Hungarica*, XLIV, 271–281.

Porto, W. and Pérez-Gonzalès, A. (2019). Redescription of the New-Zealand harvestman *Nuncia obesa obesa* (Opiliones: Lianatores: Triaenonychidae) and implications for the supposed transcontinental distribution of Nuncia. *Journal of Arachnology*, 47, 370–376.

Prat, P., Charrier, M., Deleporte, S., and Frenot, Y. (2002). Digestive carbohydrases in two epigeic earthworm species of the Kerguelen Islands (sub-Antarctic). *Pedobiologia*, 46, 417–427.

Preece, R.C. (2001). Introduced land molluscs on the islands of the Tristan da Cunha – Gough group (South Atlantic). *Journal of Conchology*, 37, 253–259.

Pugh, P.J.A. (2004). Biogeography of spiders (Araneae: Arachnida) on the islands of the Southern Ocean. *Journal of Natural History*, 38, 1461–1487.

Pugh, P.J.A. and Smith, R.I.L. (2011). *Notodiscus* (Charopidae) on South Georgia: Some implications of shell size, shell shape, and site isolation in a singular sub-Antarctic land snail. *Antarctic Science*, 23, 442–448.

Quinlan, J. (1967). The brachypterous genera and species of Eucoilidae (Hymenoptera), with description and figures of some species. *Proceedings of the Royal Entomological Society of London. Series B: Taxonomy*, 36, 1–10.

Rapp, M. (2013). Diversité, écologie et répartition des Invertébrés de l'Île de la Possession. Report, TAAF, 42.

Remaudière, G. and Etienne, J. (1988). Les Aphididae (Hom.) des îles et archipels de l'Océan Indien. *L'agronomie tropicale*, 43, 327–346.

Renault, D., Chevrier, M., Laparie, M., Vernon, P., and Lebouvier, M. (2015). Characterization of the habitats colonized by the alien Ground Beetle *Merizodus soledadinus* at the Kerguelen Islands. *Revue d'Ecologie (Terre et Vie)*, Suppt 12, 28–32.

Renault, D., Lombard, M., Vingère, J., and Laparie, M. (2016). Comparative salinity tolerance in native flies from the sub-Antarctic Kerguelen Islands: A metabolomic approach. *Polar Biology*, 39, 47–56.

Roderick, G. and Vernon, P. (2009). Invasion biology. In *Encyclopedia of Islands*, Gillespie, R.G. and Clague, D.A. (eds). University of California Press, Berkeley.

Ross, J.C. (1847). *A Voyage of Discovery and Research in the Southern and Antarctic Regions During the Years 1839–43*. John Murray, London.

Saether, O.A. and Andersen, T. (2011). Chironomidae from Gough, Nightingale and Tristan da Cunha islands. *Zootaxa*, (2915), 1–19.

Séguy, E. (1940). IV Diptères. In *Croisière du Bougainville aux îles australes françaises. Mémoires du Muséum National d'Histoire Naturelle, XIV*, Jeannel, R. (ed.). Muséum National d'Histoire Naturelle, Paris.

Séguy, E. (1953). Insectes mallophages, anoploures et diptères recueillis par M. Paulian aux Îles Kerguelen. *Mémoires de l'Institut Scientifique de Madagascar*, E(IV), 553–615.

Séguy, E. (1965a). Deux nouveaux Tendipédides des Îles Crozet (*Insectes Diptères Nématocères*). *Bulletin du Muséum National d'Histoire Naturelle*, 37, 285–289.

Séguy, E. (1965b). Morphologie du *Listriomastax litorea* End. (Insecte Diptère Coelopide). *Bulletin du Muséum National d'Histoire Naturelle*, 37, 139–144.

Séguy, E. (1971). Diptera. In *Marion and Prince Edward Islands. Report on the South African Bioogical and Geological Expedition 1965–1966*, van Zinderen Bakker Sr., E.M., Winterbottom, J.M., and Dyer, R.A. (eds). A.A. Balkema, Cape Town, South Africa.

Serra-Tosio, B. (1982). Description du mâle de *Belgica albipes* (Séguy, 1965), n. comb., rare chironomidé microptère des Îles Crozet (Diptera). *Revue Française d'Entomologie*, 4, 97–100.

Serra-Tosio, B. (1986). Un nouveau chironomide antarctique des Îles Crozet, *Parochlus crozetensis* n. sp. (*Diptera, Nematocera*). *Nouvelle Revue d'Entomologie*, 3, 149–159.

Serra-Tosio, B. and Brundin, L. (1990). Redescription of the male of *Microzetia mirabilis* Séguy, 1965, an endemic chrironomid midge from Crozet Islands (Diptera, Chironomidae). *Annales de la Société Entomologique de France (N.S.)*, 26, 411–419.

Smith, K.G.V. (2013). An introduction to the immature stages of British flies. *Handbooks for the Identification of British Insects*, 10, 163 pp.

Smithers, C.N. (1974). Arthropoda of the sub-Antarctic islands of New Zealand (4) Psocoptera. *Journal of the Royal Society of New Zealand*, 4, 315–318.

Soubeyran, Y., Meyer, J.Y., Lebouvier, M., De Thoisy, B., Lavergne, C., Urtizberea, F., and Kirchner, F. (2015). Dealing with invasive alien species in the French overseas territories: Results and benefits of a 7-year Initiative. *Biological Invasions*, 17, 545–554.

Steel, W.O. (1970). Coleoptera: Staphylinidae of South Georgia. *Pacific Insects Monograph*, 23, 240–242.

Steffan, W.A. (1972). *Lycoriella solani* (Winnetrz) from the Crozet Islands (Diptera, Sciaridae). *Pacific Insects*, 14, 429–431.

Stuke, J.H. and Bächli, G. (2015). Faunistical data of Carnidae (Diptera) from Switzerland and additional countries with the description of three new *Meoneura* species. *Mitteilungen der Schweizerischen Entomologischen Gesellschaft*, 88, 379–401.

Svanella-Dumas, L., Candresse, T., Hullé, M., and Marais, A. (2013). Distribution of Barley yellow dwarf virus-PAV in the sub-Antarctic Kerguelen Islands and characterization of two new *Luteovirus* Species. *PLoS ONE*, 8, e67231.

Tétry, A. (1947). Croisière du Bougainville aux îles australes françaises. XVIII. Oligochètes. *Mémoires du Muséum National d'Histoire Naturelle*, XX, 101–110.

Tillyard, M.A. (1920). The insects of Macquarie Island. In *Australian Antractic Expedition 1911–1914*, Mawson, D. (ed.). Government Printer, Sydney.

Tréhen, P. (1982). Cycles de développement et stratégies de la reproduction chez quelques espèces de diptères des îles subantarctiques. *Colloque sur les écosystèmes subantarctiques, CNFRA (Comité National Français des Recherches Antarctiques)*, 51, 149–156.

Tréhen, P. and Delettre, Y. (1976). Aspects originaux de la biologie et l'écologie de *Limnophyes pusillus* Eaton à Kerguelen. *Bulletin du CNFRA*, 41, 73–84.

Tréhen, P. and Vernon, P. (1982). Peuplement diptérologique d'une île subantarctique : la Possession (46°S, 51°E; Iles Crozet). *Revue d'écologie et de biologie du sol*, 19, 105–120.

Tréhen, P. and Voisin, J.F. (1984). Sur la présence de *Merizodus soledadinus* Guérin à Kerguelen (Coléoptères Trechidae). *L'Entomologiste*, 53–54.

Tréhen, P., Bouché, M., Vernon, P., and Frenot, Y. (1985). Organization and dynamics of Oligochaeta and Diptera on Possession Island. In *Antarctic Nutrient Cycles and Food Webs*, Siegfried, W.R., Condy, P.R., and Laws, R.M. (eds). Springer Verlag, Berlin.

Tréhen, P., Vernon, P., Delettre, Y., and Frenot, Y. (1986). Organisation et dynamique des peuplements diptérologiques à Kerguelen. Mise en évidence de modifications liées à l'insularité. *Bulletin du CNFRA*, 58, 241–253.

Valera, F., Martín-Vivaldi, M., and Carles-Tolrá, M. (2006). Life-history variation in three coexisting species of carnid flies (Diptera: Carnidae), *Carnus hemapterus*, *Hemeromya anthracina* and *Hemeromya longirostris*. *European Journal of Entomology*, 103, 347–353.

Van Der Putten, N., Verbruggen, C., Ochyra, R., Verleyen, E., and Frenot, Y. (2010). Sub-Antarctic flowering plants: Pre-glacial survivors or post-glacial immigrants? *Journal of Biogeography*, 37, 582–592.

Vernon, P. (1981). Peuplement diptérologique des substrats enrichis en milieu insulaire subantarctique (Îles Crozet) : étude des Sphaeroceridae du genre *Anatalanta*. Thesis, University of Rennes 1, Rennes, France.

Vernon, P. (1986). Evolution des réserves lipidiques en fonction de l'état physiologique des adultes dans une population expérimentale d'un diptère subantarctique : *Anatalanta aptera* Eaton (Sphaeroceridae). *Bulletin de la Société d'Ecophysiologie*, 11, 95–116.

Vernon, P. and Vannier, G. (1996). Developmental patterns of supercooling capacity in a sub-Antarctic wingless fly. *Experientia*, 52, 155–158.

Vernon, P. and Voisin, J.F. (1990). Faune entomologique de la Grande Ile des Apôtres (archipel Crozet, Océan Indien Austral). *Bulletin de la Société Entomologique de France*, 95, 263–268.

Vernon, P., Vannier, G., and Trehen, P. (1998). A comparative approach to the entomological diversity of polar regions. *Acta Oecologica*, 19, 303–308.

Vernon, P., Caron, F., and Davies, L. (1999). Annual activity of two endemic beetles (Carabidae) at the edge between fell-field and moorland on a sub-Antarctic island. *European Journal of Soil Biology*, 35, 39–43.

Verrall, M.C. (1879). Zoology of Kerguelen Island. Diptera. *Philosophical Transactions of the Royal Society of London*, 68, 238–248.

Viette, P. (1948). Croisière du Bougainville aux îles australes françaises. XX. Lépidoptères. *Mémoires du Muséum National d'Histoire Naturelle*, 27, 1–28.

Viette, P. (1968). Descriptions de nouvelles espèces de microlépidoptères de Madagascar et de l'Ile Marion. *Bulletin mensuel de la Société Linnéenne de Lyon*, 37, 83–91.

Vincent, R. and Voisin, J.F. (1991). *Aridionomus nodifer* Westwood dans l'archipel Crozet (Col. Lathridiidae). *L'Entomologiste*, 47, 170.

Vink, C. and Kean, J. (2013). PCR gut analysis reveals that *Tenuiphantes tenuis* (Araneae: Linyphiidae) is a potentially significant predator of Argentine stem weevil, *Listronotus bonariensis* (Coleoptera: Curculionidae) in New Zealand pastures. *New Zealand Journal of Zoology*, 40, 304–313.

Vitali Di Castri, V. (1968). *Austrochtonius insularis*, nouvelle espèce de pseudoscorpions de l'archipel de Crozet (Heterosphyronida, Chthoniidae). *Bulletin du Muséum National d'Histoire Naturelle, 2ème série*, 40, 141–148.

Vogel, M. and Nicolai, V. (1983). Invertebrates collected at the old whaling station, Grytviken, South Georgia. *Polar Record*, 135, 607–609.

Voisin, J.F. (1975). *Vanessa cardui* (L.) dans l'archipel Crozet. *Bulletin de la Société Entomologique de France*, 80, 80–81.

Voisin, J.F. (1976). Sur le genre *Bothrometopus* Jeannel 1940 (Col., Curculionidae Ectemnorchininae). *Bulletin de la Société Entomologique de France*, 81, 9–16.

Voisin, J.F. (1986). Sur la présence de *Pseudeuplectus antarcticus* Enderlein à l'Île aux Cochons (archipel Crozet) (Col. Pselaphidae). *L'Entomologiste*, 42, 151–152.

Voisin, J.F., Chapelin-Viscardi, J.D., Ponel, P., and Rapp, M. (2015). *Xanium vanhoeffenianum* Enderlein, 1904 et *X. possessionense* Hoffmmann, 1966 : espèces distinctes de l'archipel Crozet (Coleoptera, Curculionidae). *Bulletin de la Société Entomologique de France*, 120, 431–434.

Voisin, J.F., Chapelin-Viscardi, J.D., Ponel, P., and Rapp, M. (2017). *Les Coléoptères de la province de Kerguelen (îles subantarctiques de l'océan Indien). Faune de France 99*, Fédération Française des Sociétés de Sciences Naturelles, Paris.

Volonterio, O., De Leon, R.P., Convey, P., and Krzeminska, E. (2013). First record of Trichoceridae (Diptera) in the maritime Antarctic. *Polar Biology*, 36, 1125–1131.

Watson, K.C. (1967). The terrestrial arthropods of Macquarie island. *ANARE Scientific Reports*, 99, 21–23.

Watt, J.C. (1971). Entomology of the Aucklands and other islands south of New Zealand: Coleoptera: Scarabaeidae, Byrrhidae, Ptinidae, Tenebrionidae. *Pacific Insect Monographs*, 27, 193–224.

Wearing-Wilde, J. (1996). Mate choice and competition in the barklouse *Lepinotus patruelis* (Psocoptera: Trogiidae): The effect of diet quality and sex ratio. *Journal of Insect Behavior*, 9, 599–612.

Wilson, T.H. and Stannard, L.J. (1970). Thysanoptera of South Georgia. *Pacific Insects Monograph*, 23, 221–226.

Wise, K.a.J. (1964). Insects of Campbell Island. Coleoptera, Ptinidae. *Pacific Insect Monographs*, 7, 395–396.

Womersley, H. (1937). Diptera. Family Simuliidae. *British, Australian and New Zealand Antarctic Research Expedition 1929–1931 Report, Series B.*

Wyville Thomson, C. and Murray, J. (1885). Report on the scientific results of the voyage of H.M.S. Challenger during the years 1873–76. Report, Majesty Stationary Office, London.

Ysnel, F. and Ledoux, J.C. (1988). Données sur le cycle biologique de quelques araignées des terres Australes Françaises (Kerguelen, Crozet). *Bulletin de la Société Scientifique de Bretagne*, 59, 209–221.

Zur Strassen, R. (1982). On Thysanoptera (Insecta) from some islands in the sub-Antarctic part of the Indian Ocean. *Colloque sur les écosystèmes subantarctiques, CNFRA (Comité National Français des Recherches Antarctiques)*, 48, 159–169.

Index

D, E, F

H, K, L

M, N, O

P, R, S

T, V, X

Other titles from

ISTE

in

Biology

2021

GRANDCOLAS Philippe, MAUREL Marie-Christine
Systematics and the Exploration of Life

2020

CAZEAU Cyrille
Foot Surgery Viewed Through the Prism of Comparative Anatomy: From Normal to Useful (Comparative Anatomy and Posture of Animal and Human Set – Volume 1)

DUJON Bernard, PELLETIER Georges
Trajectories of Genetics

2019

BRAND Gérard
Discovering Odors

BUIS Roger
Biology and Mathematics: History and Challenges

www.ingramcontent.com/pod-product-compliance
Lightning Source LLC
LaVergne TN
LVHW011801200926
848484LV00003B/750